CONTINUOUS-TIME SIGMA-DELTA MODULATION
FOR A/D CONVERSION IN RADIO RECEIVERS

THE KLUWER INTERNATIONAL SERIES
IN ENGINEERING AND COMPUTER SCIENCE

ANALOG CIRCUITS AND SIGNAL PROCESSING
Consulting Editor: **Mohammed Ismail**. *Ohio State University*

Related Titles:

DIRECT DIGITAL SYNTHESIZERS: THEORY, DESIGN AND APPLICATIONS
  J. Vankka, K. Halonen
  ISBN: 0-7923 7366-9
SYSTEMATIC DESIGN FOR OPTIMISATION OF PIPELINED ADCs
  J. Goes, J.C. Vital, J. Franca
  ISBN: 0-7923-7291-3
OPERATIONAL AMPLIFIERS: Theory and Design
  J. Huijsing
  ISBN: 0-7923-7284-0
HIGH-PERFORMANCE HARMONIC OSCILLATORS AND BANDGAP REFERENCES
  A. van Staveren, C.J.M. Verhoeven, A.H.M. van Roermund
  ISBN: 0-7923-7283-2
HIGH SPEED A/D CONVERTERS: Understanding Data Converters Through SPICE
  A. Moscovici
  ISBN: 0-7923-7276-X
ANALOG TEST SIGNAL GENERATION USING PERIODIC ΣΔ-ENCODED DATA STREAMS
  B. Dufort, G.W. Roberts
  ISBN: 0-7923-7211-5
HIGH-ACCURACY CMOS SMART TEMPERATURE SENSORS
  A. Bakker, J. Huijsing
  ISBN: 0-7923-7217-4
DESIGN, SIMULATION AND APPLICATIONS OF INDUCTORS AND TRANSFORMERS FOR Si RF ICs
  A.M. Niknejad, R.G. Meyer
  ISBN: 0-7923-7986-1
SWITCHED-CURRENT SIGNAL PROCESSING AND A/D CONVERSION CIRCUITS: DESIGN AND IMPLEMENTATION
  B.E. Jonsson
  ISBN: 0-7923-7871-7
RESEARCH PERSPECTIVES ON DYNAMIC TRANSLINEAR AND LOG-DOMAIN CIRCUITS
  W.A. Serdijn, J. Mulder
  ISBN: 0-7923-7811-3
CMOS DATA CONVERTERS FOR COMMUNICATIONS
  M. Gustavsson, J. Wikner, N. Tan
  ISBN: 0-7923-7780-X
DESIGN AND ANALYSIS OF INTEGRATOR-BASED LOG -DOMAIN FILTER CIRCUITS
  G.W. Roberts, V. W. Leung
  ISBN: 0-7923-8699-X
VISION CHIPS
  A. Moini
  ISBN: 0-7923-8664-7
COMPACT LOW-VOLTAGE AND HIGH-SPEED CMOS, BiCMOS AND BIPOLAR OPERATIONAL AMPLIFIERS
  K-J. de Langen, J. Huijsing
  ISBN: 0-7923-8623-X

CONTINUOUS-TIME SIGMA-DELTA MODULATION FOR A/D CONVERSION IN RADIO RECEIVERS

by

Lucien Breems
Philips Research, Eindhoven, The Netherlands

and

Johan H. Huijsing
Delft University of Technology, The Netherlands

KLUWER ACADEMIC PUBLISHERS
BOSTON / DORDRECHT / LONDON

A C.I.P. Catalogue record for this book is available from the Library of Congress.

ISBN 0-7923-7492-4

Published by Kluwer Academic Publishers,
P.O. Box 17, 3300 AA Dordrecht, The Netherlands.

Sold and distributed in North, Central and South America
by Kluwer Academic Publishers,
101 Philip Drive, Norwell, MA 02061, U.S.A.

In all other countries, sold and distributed
by Kluwer Academic Publishers,
P.O. Box 322, 3300 AH Dordrecht, The Netherlands.

Printed on acid-free paper

Printed in the Netherlands.

Table of contents

List of abbreviations   ix

List of symbols   xi

Preface   xiii

1   Introduction   **1**

    1.1   The world of communication 1
    1.2   Sigma-delta A/D conversion 3
    1.3   System level simulation 5
    1.4   Motivation and objectives 6
    1.5   Organization of the work 6
        References 8

2   A/D conversion in radio receivers   **9**

    2.1   Introduction 9
    2.2   From baseband to RF A/D conversion 10
        2.2.1   Heterodyne receiver with baseband A/D conversion 11
        2.2.2   Heterodyne receiver with IF digitizing 13
        2.2.3   Receiver architecture with RF digitizing 14
    2.3   Sigma-delta modulation in a heterodyne receiver 14
    2.4   Performance parameters 16
        2.4.1   Dynamic range 16
        2.4.2   Linearity 17
        2.4.3   Image rejection 19
        2.4.4   Figure-of-merit 21
    2.5   GSM and AM/FM radio specifications 22
    2.6   Summary 24
        References 26

3 Continuous-time sigma-delta modulation 29

3.1 Introduction 29
3.2 Theory of sigma-delta modulation 30
    3.2.1 Oversampling and noise-shaping 30
    3.2.2 Tones 33
    3.2.3 Harmonic distortion 37
    3.2.4 Intersymbol interference 41
    3.2.5 Phase jitter 43
    3.2.6 Aliasing 44
3.3 Linear stability analysis 46
    3.3.1 Small signal stability 46
    3.3.2 Large signal stability 51
3.4 High-frequency stability compensation 53
    3.4.1 Feedforward compensation 53
    3.4.2 Feedforward compensation and local feedback 59
    3.4.3 Feedback compensation 60
    3.4.4 Feedback compensation and local feedback 61
    3.4.5 Feedforward versus feedback compensation 63
3.5 Quadrature sigma-delta modulation 64
    3.5.1 Complex integrator 64
    3.5.2 Complex filter design 66
    References 70

4 Realization of an IF-to-baseband sigma-delta modulator 73

4.1 Introduction 73
4.2 Frequency translation in sigma-delta modulators 74
    4.2.1 Mixer inside the sigma-delta loop 74
    4.2.2 Mixer outside the sigma-delta loop 77
    4.2.3 Mixer inside loop versus outside loop 78
4.3 IF mixer design 79
    4.3.1 Mixer topology 79
    4.3.2 Linearity performance 80
    4.3.3 Local oscillator driver 82
    4.3.4 Self-mixing 88
4.4 IF sigma-delta modulator design 90
    4.4.1 IF sigma-delta modulator topology 90
    4.4.2 Input filter stage design 92

4.4.3 Transconductance-C integrator 96
4.4.4 A/D and D/A converter 97
4.4.5 LO driver scheme 98
4.5 Experimental results 99
4.5.1 Test chip 1: Baseband sigma-delta modulator 100
4.5.2 Test chip 2: IF-to-baseband sigma-delta modulator 102
4.5.3 Performance summary 105
4.6 Conclusions 106
References 108

5 Realization of a quadrature sigma-delta modulator 109

5.1 Introduction 109
5.2 Image interference 110
5.3 Dynamic element matching 111
5.3.1 Complex data-controlled DEM algorithm 117
5.3.2 DEM implementation 118
5.3.3 Impedance mismatch of DEM and DAC switches 123
5.3.4 Charge injection of DEM and DAC switches 125
5.4 Quadrature sigma-delta modulator design 129
5.4.1 Topology 129
5.4.2 Input stage 131
5.4.3 Feedback and DEM circuit 132
5.5 Experimental results 135
5.6 Conclusions 139
References 141

6 Benchmark 143

6.1 Introduction 143
6.2 Benchmark of test chips 144
6.3 Conclusions 147
References 149

Index 153

List of abbreviations

A/D	analog-to-digital
ADC	analog-to-digital converter
AGC	automatic gain control
AM	amplitude modulation
BER	bit error rate
BW	bandwidth
CLK	clock
CM	cross-modulation
CT	continuous-time
D/A	digital-to-analog
DAC	digital-to-analog converter
DEM	dynamic element matching
DLL	delay-locked loop
DR	dynamic range
DSP	digital signal processor
EXOR	exclusive-or
FFT	fast fourier transform
FM	frequency modulation
FOM	figure-of-merit
GMSK	gaussian minimum-shift-keying
HD	harmonic distortion
I	in-phase
IF	intermediate frequency
IM	intermodulation distortion
IP	intercept point
IR	image rejection
LNA	low-noise amplifier
LO	local oscillator
LSB	least significant bit
NTF	noise transfer function
NZIF	near-zero-IF
OPAMP	operational amplifier
OSR	oversampling ratio
P	power

List of abbreviations

PLL	phase-locked loop
Q	quadrature-phase
RF	radio frequency
RTZ	return-to-zero
SAW	surface acoustic wave
STF	signal transfer function
$\Sigma\Delta$	sigma-delta
SFDR	spurious-free dynamic range
SINAD	signal-to-noise-and-distortion
SNDR	signal-to-noise-and-distortion ratio
SNR	signal-to-noise ratio
SQNR	signal-to-quantization-noise ratio
V/I	voltage-to-current
ZIF	zero-IF

List of symbols

Δ	relative error	-
Δa	relative gain error	-
$\Delta\varphi$	phase error	rad,°
Δ_q	quantization step size	-
δ	duty cycle	-
φ	phase	rad,°
κ	quantizer gain	-
κ_s	minimum (large signal) stable quantizer gain	-
μ	charge carrier mobility	cm^2/Vs
π	pi, 3.141593	-
θ	quantizer phase	°
σ_j	standard deviation	-
τ	time constant	s
ω_u	unity gain frequency	rad/s
ω_z	zero frequency	rad/s
A	DC gain	-
a_n	nth butterworth coefficient	-
b_n	nth feedback coefficient	-
C	symbol for capacitor	F
C_{db}	drain-bulk capacitor	F
C_{ds}	drain-source capacitor	F
C_{gd}	gate-drain capacitor	F
C_{gs}	gate-source capacitor	F
C_i	integration capacitor	F
C_{ox}	normalized oxide capacitance	F
c_n	nth feedforward coefficient	-
d_n	nth local feedback coefficient	-
e	quantization error	-
f,ω	frequency	Hz, rad/s
f_b	bandwidth	Hz
f_{LO}	local oscillator frequency	Hz
f_s	sampling frequency	Hz
g_m	transconductance	S
$H(s)$	loopfilter transfer function	-
I_b	bias current	A

List of symbols

I_{os}	offset current	A
I_{dac}	DAC current	A
i	small signal current	A
i_{dac}	transient DAC current	A
i_{in}	transient input current	A
i_r	residue current	A
j	complex operator	-
k	Boltzmanns' constant, $1.3805 \cdot 10^{-23}$	J/K
L	channel length of MOS transistor	μm
m	oversampling ratio	-
Nj	jitter power	W
N_q	quantization noise power	W
N_{th}	thermal noise power	W
P	power	W
P_{in}	input power	W
Q	charge	C
q	quantization level	-
RTZ	return-to-zero duty cycle	-
R_{dac}	DAC resistor	Ω
R_{in}	input resistor	Ω
R_L	load resistor	Ω
R_s	degeneration resistor	Ω
r_{on}	switch ON-impedance	Ω
s	Laplace operator	rad/s
T	temperature	K
T_s	sampling period	s
t	time	s
V_{DD}	positive supply voltage of MOS circuits	V
V_{SS}	negative supply voltage of MOS circuits	V
V_{dac}	DAC voltage	V
v_{dac}	transient DAC voltage	V
V_{ds}	drain-source voltage	V
V_g	gate-to-ground voltage of MOS transistor	V
V_{gs}	gate-source voltage	V
$V_{gs,eff}$	effective gate-source voltage	V
V_{in}	input voltage	V
v_{in}	transient input voltage	V
V_{os}	offset voltage	V
v_r	residue voltage	V
V_{Th}	threshold voltage	V
W	channel width of MOS transistor	μm

Preface

This book describes the theory, design and realizations of continuous-time $\Sigma\Delta$ modulators for analog-to-digital conversion in radio receivers. The challenge of the work is the design of a $\Sigma\Delta$ modulator with high linearity, large dynamic range and strong image rejection capabilities. With such an A/D converter, requirements for a receiver architecture in terms of selectivity and sensitivity can be relaxed, resulting in a cheaper system with a higher level of integration.

Important trends in the receiver design for wireless portable applications are: smaller product sizes, cheaper products and longer stand-by times. Products can be made smaller and cheaper by increasing the level of integration. This means on-chip integration of external components, such as inductors and filters. Herein, an important role can be played by the A/D converter. Shifting the A/D converter towards the antenna side of the receiver, allows more digital integration of (external) analog functions on a single digital chip. However, this requires an A/D converter with high linearity, dynamic range, bandwidth and image rejection capabilities.

In chapter 2, it will be shown that the required performance of the ADC depends very much on its place in the receiver architecture. Single-bit continuous-time $\Sigma\Delta$ modulation is a good technique for A/D conversion in receivers, as it incorporates inherent anti-aliasing filtering, excellent linearity performance, and low-power capability. All these performance aspects are particularly important in battery-powered receivers. The main performance parameters are described and a figure-of-merit is presented that is used for comparison between different designs.

In chapter 3, the theory of higher-order continuous-time $\Sigma\Delta$ modulation is described. Important aspects, determining the performance of a continuous-time $\Sigma\Delta$ modulator, are quantization noise, DC tones, intersymbol interference, clock jitter, and aliasing. The design of higher-order filters is described, with Butterworth and inverse-Chebyshev filter characteristics. With the availability of quadrature signals in a radio receiver, the theory of quadrature $\Sigma\Delta$ modulation is treated as well.

Chip realizations are described in chapter 4 and chapter 5. Chapter 4 describes the design of a $\Sigma\Delta$ modulator with integrated mixer. The key features of this IF-to-baseband A/D converter are the high linearity and low-power consumption. In chapter 5 the design of a quadrature $\Sigma\Delta$ modulator with a data-dependent dynamic element matching circuit is shown. Measurements on this A/D converter show a high image rejection of 63 dB typically.

In the last chapter the main conclusions are summarized. The performance of the prototype test chips that have been described in this book are compared with other state-of-the-art test designs from literature.

The authors wish the reader a pleasant time in investigating the interesting aspects of continuous-time $\Sigma\Delta$ modulation for A/D conversion in radio receivers.

Lucien J. Breems
Johan H. Huijsing

Eindhoven, May 2001

Introduction

1

This book describes the theory, design and realization of continuous-time sigma-delta ($\Sigma\Delta$) modulators for analog-to-digital (A/D) conversion in radio receivers. The challenge of the work is the design of a $\Sigma\Delta$ modulator with high linearity, large dynamic range and strong image rejection capabilities. With such an A/D converter (ADC), requirements for a receiver architecture in terms of selectivity and sensitivity can be relaxed, resulting in a cheaper system with a higher level of integration. In this introduction, general aspects regarding receiver design and sigma-delta A/D conversion are described. In addition, the main challenges, motivation and organization of this book are presented.

1.1 The world of communication

In today's world, the market for information and communication technology is expanding as never before. In the last century, the communication market grew from zero to about 1 billion telephone connections worldwide [1.1]. It is to be expected that this amount will double in the next twenty years (Fig.1-1). Not surprisingly, wireless communication has become more and more important. Nowadays, in a modern luxury car, besides a radio, one can also find a cellular phone, fax, GPS receiver, television or even an internet browser. All these portable wireless communication systems have receivers to retrieve the information signals from the outside world. The signals are received by an antenna, and the desired signal band, for example a GSM channel, is selected from the total received spectrum. This frequency band undergoes analog filtering, amplification, frequency modulation and analog-to-digital conversion. Further signal processing is done in the digital domain by a digital signal processor (DSP).

Important trends in the receiver design for wireless portable applications are: smaller product sizes, cheaper products and longer stand-by times. Products can be made smaller and cheaper by increasing the level of integration. This means on-chip integration of external components, such as inductors and filters. For example, in the last eight years, the total component count for a GSM RF front-end has reduced from 500-600 components down to only 150 [1.2]. Herein, an important role can be played by the A/D converter. Shifting the A/D converter towards the antenna side of the receiver, allows more digital integration of (external) analog functions on a single digital chip. However, this requires an A/D converter with high linearity, dynamic range and bandwidth capabilities (chapter 2), which conflicts with the low-power requirements for a long stand-by time; A great challenge for mixed-signal designers!

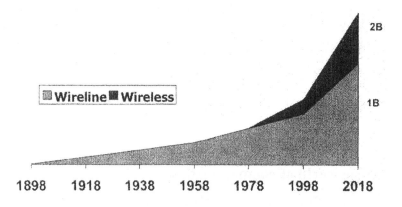

Fig. 1-1   *Total number of wireline and wireless phone connections worldwide [1.1] (courtesy of P.P. 't Hoen)*

1.2 Sigma-delta A/D conversion

Signals that are progressing continuously in time and amplitude are classified as analog signals. An analog-to-digital converter produces a digital representation of the analog input signal by sampling the input signal at discrete time moments and quantizing the amplitude of the input signal in discrete amplitude levels. Due to the finite number of quantization levels, the quantization process causes errors, which set the maximum achievable resolution. The resolution of the ADC can be improved by increasing the number of quantization levels. In a sigma-delta modulator, additional techniques are used to achieve higher accuracy, namely oversampling and noise-shaping. *Oversampling* means that sampling of the analog input signal is done with a sampling rate higher than the minimum required Nyquist frequency, which is twice the signal bandwidth. Accordingly, the oversampling ratio (OSR) is defined as

$$m = \frac{f_s}{2f_b} \qquad (1\text{-}1)$$

where m is the oversampling ratio, f_s is the sampling rate and f_b is the signal bandwidth. *Noise-shaping* implies filtering of the quantization errors, in order to shape their frequency response. As a result, the quantization error power is reduced in the frequency band of interest, while it is increased outside that band. This way, high resolution can be obtained in a relatively small bandwidth. The general model of a single-loop sigma-delta modulator is shown in Fig. 1-2a [1.3]. Basically, a sigma-delta modulator consists of a loopfilter, performing the noise-shaping, a low resolution quantizer, which is oversampled, and a feedback loop. The loopfilter can be a lowpass or bandpass filter. A first-order lowpass filter is an accumulator in the discrete-time domain or an integrator in the continuous-time domain. The quantizer can be modeled as a summation node for the loopfilter output signal and the quantization error E_n (Fig. 1-2b).

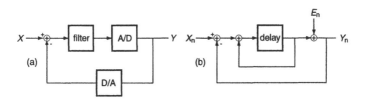

Fig. 1-2 *General model of a $\Sigma\Delta$ modulator (a); Discrete-time model of a first-order lowpass $\Sigma\Delta$ modulator (b)*

The delay cell with the positive feedback path is the mathematical model of an accumulator. The output Y of the sigma-delta modulator of Fig. 1-2b is equal to

$$Y_n = X_{n-1} - (E_n - E_{n-1}) \qquad (1\text{-}2)$$

where X is the input signal. It can be observed from Eq. (1-2) that the modulator passes the input signal unchanged, with one sampling period delay, while the quantization error is differentiated. For highly oversampled (slow varying) input signals, the average of the quantization errors is close to zero. For high-frequency input signals, the quantization error becomes large. Fig. 1-3 shows a typical frequency spectrum of the output Y of a first-order sigma-delta modulator. The horizontal axis represents the (relative) frequency axis, and the vertical axis shows the signal amplitude in decibels (dB). The frequency of the input signal is 1000 times lower than the sampling rate (500 times oversampled) and can be observed at spectral position 10^{-3}. The quantization errors of a first-order $\Sigma\Delta$ modulator are discrete tones in the frequency spectrum. The power of the tones is small at low frequencies and large around half the sampling frequency. The high-frequency quantization noise is filtered out by a decimation filter [1.4], [1.5] behind the modulator. More effective noise-shaping is achieved with a higher-order $\Sigma\Delta$ modulator which has more accumulator/integrator stages. However, due to the feedback loop, higher-order modulators suffer from stability problems, and frequency compensation is needed to stabilize the loop (chapter 3).

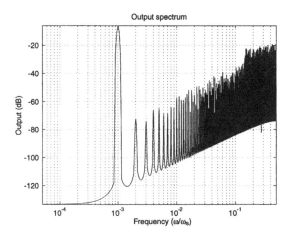

Fig. 1-3   *Simulated output frequency spectrum of a first-order single-bit $\Sigma\Delta$ modulator (input signal at 10^{-3})*

1.3 System level simulation

A powerful method to quickly predict and validate the performance of a $\Sigma\Delta$ modulator is simulation of a numerical model at the system level. In this book, numerous system level simulations have been performed, with certain non-ideal behavior included. This way, the effect of non-idealities can be investigated in a matter of minutes, without the need to go through the whole design process. It will be shown in chapters 4 and 5 by means of chip realizations and experimental results, that system level simulations are very useful to predict the effect of non-idealities. The system level simulations have been done with a software tool, which has been developed with MATLAB®. A screenshot of the graphical user interface is shown in Fig. 1-4. The tool consists of an input panel for the Nyquist bandwidth, oversampling factor, input frequency, amplitude, data filter, and other parameters. With these input parameters, the model of the $\Sigma\Delta$ modulator is simulated over a number of periods. The output bitstream is filtered and the frequency spectrum is shown in the result window. Performance parameters, such as signal-to-noise ratio, harmonic distortion, and image rejection can be extracted directly from the frequency spectrum.

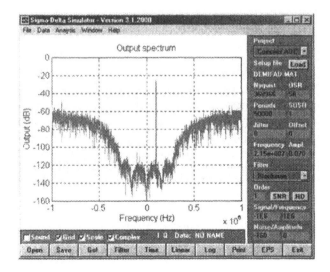

Fig. 1-4 *GUI of software tool for system level $\Sigma\Delta$ simulation*

1.4 Motivation and objectives

The main *practical* motivation for the work in this book, is the demand for *high-performance* and *low-power* A/D converters in receivers. With a high performance A/D converter, the level of integration of a receiver can be increased, for example by means of digital filters. Increasing the level of integration reduces component count, cost, and product size. This implies higher demands on the ADC concerning linearity, dynamic range, input frequency and bandwidth. In this book three main subjects have been investigated which are

- *linearity*
- *power consumption*
- *image rejection performance*

Low-power consumption is an important issue in battery-powered applications, like mobile phones. Improving the performance, while reducing the total power, requires new design topologies and innovative solutions. This is an important *scientific* motivation for exploring the relations between power and performance. In this work, the total power consumption is optimized at three levels in the top-down design process:

- *system architecture level*
- *functional block level*
- *circuit level*

This book describes the results of these three design steps. First, the place of the ADC in the total system architecture is analyzed, which has a strong relation with the required performance. Secondly, the power is reduced by combining different functional blocks (mixer, filter and ADC) and the power-to-performance relations are analyzed. The last objective is the design of prototype test chips to show that high performance can be realized at low-power consumption.

1.5 Organization of the work

In this book the theory and design of continuous-time ΣΔ modulators for A/D conversion in radio receivers is described. In chapter 2, it will be shown that the required performance of the ADC depends very much on its place in the receiver. Single-bit continuous-time ΣΔ modulation is a good technique for A/D conversion in receivers, as it incorporates inherent anti-aliasing filtering, excellent linearity performance, and low-power capability. All these performance aspects are particularly important in battery-powered receivers. The main performance parameters are described and a figure-of-merit is presented that can be used for comparison between different designs. Moreover, A/D converter specifications for GSM mobile phone and AM/FM radio are presented, which are the main applications of this work.

In chapter 3, the theory of higher-order continuous-time $\Sigma\Delta$ modulation is described. Important aspects, determining the performance of a continuous-time $\Sigma\Delta$ modulator, are quantization noise, DC tones, intersymbol interference, clock jitter, and aliasing. The design of higher-order filters is described, with Butterworth and inverse-Chebyshev filter characteristics. With the availability of quadrature signals in a radio receiver, the theory of quadrature $\Sigma\Delta$ modulation is treated as well.

Chapter 4 presents the theory, design, and realization of a continuous-time $\Sigma\Delta$ modulator with integrated mixer for A/D conversion of IF signals. The work focuses on high linearity performance and low-power consumption of the modulator. Theoretical results and measured performance of the prototype chips are analysed.

The design of a quadrature $\Sigma\Delta$ modulator can be found in chapter 5. The modulator consists of two IF $\Sigma\Delta$ modulators as presented in chapter 4. The focus of this work is the image rejection performance. To improve image rejection performance, a novel dynamic element matching algorithm is presented, which is based on the complex output data of the quadrature $\Sigma\Delta$ modulator. Careful implementation of the dynamic element matching circuit is required, as the modulator is very sensitive to parasitic non-idealities.

In chapter 6, the main conclusions are summarized. The performance of the prototype test chips that have been described in this book are compared with other state-of-the-art test designs from literature.

References

[1.1] Hoen, P.P. 't, "Telecommunicatie: een wereldmarkt," *Tijdschrift van het nederlands elektronica- en radiogenootschap*, no. 3, pp. 117-124, 1999.

[1.2] Fenk, J. "Highly integrated RF-IC's for GSM, DECT and UMTS systems," *Proc. of ESSCIRC*, pp. 11-14, Sept. 1999.

[1.3] Norsworthy, S.R., R. Schreier, G.C. Temes, *Delta-sigma data converters - Theory, design, and simulation*, IEEE Press, New York, 1997.

[1.4] Candy, J.C., "Decimation for sigma delta modulation," *IEEE Trans. Commun.*, vol. COM-34, pp. 72-76, Jan. 1986.

[1.5] Park, S., "Multistage decimation filter design techniques for high-resolution sigma-delta A/D converters," *IEEE Trans. Instr. Measurement*, vol. 41, pp. 868-873, Dec. 1992.

A/D conversion in
radio receivers

2

2.1 Introduction

The main function of a radio receiver is the reception of a, possibly weak, desired channel from a wideband frequency spectrum containing strong interference signals, with a minimum specified signal-to-noise-and-distortion ratio. To accomplish these tasks of selectivity and sensitivity, filters and amplifiers are needed to suppress interference signals and to increase the desired channel power respectively. In general, the analog building blocks determine the sensitivity of the receiver and the filters determine the selectivity. Because the desired channel band may be modulated at a high frequency (GSM channels near 1 GHz), mixers are also used to translate the channel to more appropriate lower frequencies. The analog-to-digital converter is becoming an important part of the receiver architecture. The place of the ADC determines which functions are implemented with analog circuitry and what functionality is done in the digital signal processor. As the size of digital circuits as well as the supply voltage (and power) decreases with each new technology generation, putting more functionality into the digital signal processor (DSP) is required to take advantage of these trends. Moreover, analog signal processing functionality such as filtering and frequency translation, can be performed by the DSP with almost any degree of perfection required. In this chapter, the place of the ADC in the receiver architecture is discussed in section 2.2. In section 2.3 motivations are given for using a $\Sigma\Delta$

modulator as A/D converter in a receiver. Shifting part of the selectivity into the DSP requires an ADC with a large dynamic range and high linearity. These and other performance parameters are described in section 2.4. In section 2.5 some important issues concerning GSM mobile phone, and AM/FM radio specifications are highlighted, which are the applications of the work in this book. The conclusions of this chapter are summarized in section 2.6.

2.2   From baseband to RF A/D conversion

Basically, the architecture of a radio receiver front-end consists of an antenna, an analog signal processing part, an A/D converter, and a digital signal processor (Fig.2-1). The place of the A/D converter in a receiver front-end is of great importance as it affects overall performance, complexity, power consumption, and cost. In today's commercial receivers, the analog part includes an RF Bipolar chip e.g. for the low-noise amplifier (LNA) [2.1] and external filters, inductors, capacitors, etc. The digital part includes a single digital CMOS chip. The analog part significantly adds to the total component count which has a substantial impact on size and cost.

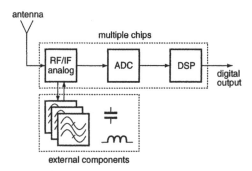

Fig. 2-1   *Partitioning of a receiver front-end*

Shifting analog functions such as filters, mixers and amplifiers into the digital domain or, in other words, moving the A/D converter towards the antenna, reduces the complexity of the receiver. However, as the A/D converter moves closer to the antenna side, the required performance specifications for the ADC become very stringent. In section 2.2.1, the traditional superheterodyne receiver with baseband A/D conversion is shown. Moving the A/D converter to the antenna side finally results in the ultimate digital receiver (section 2.2.3), however with (yet) impossible ADC specifications.

2.2.1 Heterodyne receiver with baseband A/D conversion

The model of a traditional superheterodyne receiver architecture is shown in Fig. 2-2. The antenna signal is filtered by a wideband bandpass filter and amplified by a low-noise amplifier. The desired channel, modulated at a radio frequency (RF), is mixed down to a lower intermediate frequency (IF), which for example is 10.7 MHz in FM radio applications [2.2]. External narrowband channel filters with high selectivity, such as ceramic filters or SAW filters, are available at this IF. The first local oscillator (LO) frequency (f_{LO1}) is tuned to select the desired RF channel. Before frequency translation, an image rejection (IR) filter reduces signal power at the image frequency. The channel filter passes the desired IF channel, and suppresses adjacent channels by about 30-40 dB. A variable gain amplifier with automatic gain control (AGC) fits the signal power into the dynamic input range of the subsequent blocks. Two IF mixer stages, driven by quadrature local oscillator signals ($f_{LO2,I}$ and $f_{LO2,Q}$), convert the IF channel to DC or a low frequency. Quadrature mixing of the signal to in-phase (I) and quadrature-phase (Q) channels is required to achieve a sufficient amount of image suppression (section 2.4.3). After quadrature mixing, the I and Q channels at baseband are filtered by lowpass anti-aliasing filters (section 3.2.6) and digitized. Further signal processing is done in the DSP. Requirements for the A/D converters in the superheterodyne receiver regarding dynamic range, linearity, and bandwidth are relaxed because of all the filters, in particular the channel filter, preceding the A/D converters.

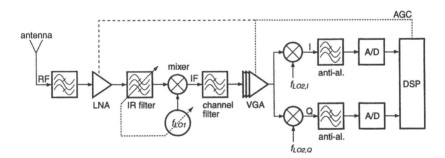

Fig. 2-2   *Traditional superheterodyne receiver architecture*

In addition, the baseband channel can be digitized at a relatively low sample rate. Therefore, power efficiency of the ADCs is high [2.3]. Image rejection requirements strongly depend on the choice of the second IF frequency of the quadrature signals. If the quadrature channel is translated to a low frequency (low-IF downconversion), the image signal and desired channel are not related, and the image signal may be stronger than the desired channel (Fig.2-3a). Therefore, image rejection has to be large and for example in the order of 80-90 dB for FM radio. The channel filter suppresses the image signal with 40 dB and another 40 dB of rejection must come from the quadrature mixing.

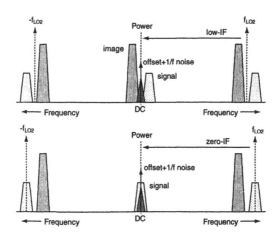

Fig. 2-3   *(a) Downconversion to low-IF; (b) Downconversion to zero-IF*

The main advantage of low-IF downconversion is that offset as well as flicker noise do not interfere with the desired signal. This is not the case if zero-IF downconversion is performed (Fig.2-3b). As the desired channel is mixed down to DC, offset and flicker noise are present in the middle of the signal band and interfere with the signal. However, in case of zero-IF downconversion, the image and the desired signal are the same. The quadrature signals are used to reconstruct the original upper and lower modulation sidebands. The required accuracy of the quadrature signals depends on the modulation technique which is used. For example, a 1 dB amplitude imbalance and 5-degree phase imbalance are acceptable for a QPSK-OFDM-QAM system [2.4].

2.2.2 Heterodyne receiver with IF digitizing

An alternative receiver architecture, which is growing more popular, is the heterodyne receiver with IF A/D conversion (Fig.2-4). The ADC is shifted in front of the IF mixers and quadrature modulation is performed by the DSP, with low-power consumption and perfect linearity and matching for excellent image rejection performance. Moreover, an IF A/D converter is insensitive to DC offset and low-frequency noise. As quadrature mixing is done in the digital domain, only one ADC is required. The channel filter in front preserves the moderate dynamic range, bandwidth, and linearity requirements for the ADC. A drawback is that the sample rate of the ADC is at least twice the IF of the input channel. Furthermore, linearity and dynamic range requirements are more difficult to meet at higher frequencies due to circuit non-idealities and parasitic effects. These constraints make the ADC for IF digitizing much less power efficient, especially if the IF is high, compared to a baseband ADC. In [2.5] - [2.9], numerous designs are proposed for bandpass A/D conversion in radio receivers.

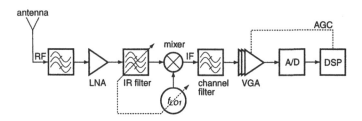

Fig. 2-4   *Heterodyne receiver architecture with IF A/D converter*

The next step is to shift the A/D converter in front of the channel selection filter (Fig.2-5) [2.10]. A wideband ADC digitizes all channels and channel selection is implemented in the DSP. This is especially advantageous in a base station for cellular phones, where only one receiver board is required with this architecture for processing all channels. In this case a higher power consumption can be allowed. Another aspect of this architecture is the flexibility in adapting to system changes, by only changing software programs [2.11]. However, the lack of analog prefiltering by a channel filter and amplification by a VGA puts heavy linearity and dynamic range requirements onto the ADC.

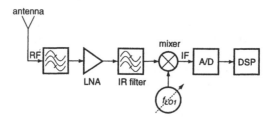

Fig. 2-5   *Receiver architecture with wideband IF A/D conversion*

2.2.3 Receiver architecture with RF digitizing

The ultimate digital receiver architecture is shown in Fig. 2-6. The A/D converter is only preceded by an RF bandpass filter and an LNA, which may have some AGC. Signal processing such as frequency translation, channel filtering, and signal demodulation is done all in one digital chip. Requirements for the A/D converter in this architecture are extremely heavy, as it has to handle the full antenna receiving power. This means the converter should have high dynamic range, high linearity, large bandwidth at high (RF) frequencies. If it were possible to design, the ADC in this architecture would most likely consume an excessive amount of power. Yet, attempts have been made to explore the boundaries of RF A/D conversion [2.12], [2.13].

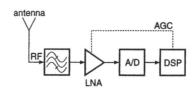

Fig. 2-6   *Receiver with RF A/D conversion*

2.3   Sigma-delta modulation in a heterodyne receiver

It was explained in section 2.2.1 that the ADCs in the traditional heterodyne receiver architecture have relaxed requirements, as the desired channel is already selected and converted to baseband. In [2.14] - [2.17] it is shown that high performance and low-power A/D converters, based on the principle of $\Sigma\Delta$ modulation, are available for baseband signal conversion. $\Sigma\Delta$ modulation is a widely used technique for high-performance and low-power A/D conversion of relatively low-bandwidth signals. Numerous designs have been presented for audio and radio applications [2.5] - [2.9], [2.14] - [2.17]. In particular, lowpass continuous-time $\Sigma\Delta$ modulation is suitable for application in the receiver architecture of Fig. 2-2, due to its low-power capability, high linearity and inherent anti-aliasing filtering (shown in the upper quadrature path in Fig. 2-7) [2.3]. These issues are discussed in detail in chapter 3. It was shown in the previous sections that digital integration of filters and mixers, in respect to a traditional receiver with baseband A/D converters, results in tough linearity, dynamic range, and bandwidth requirements. In addition, these requirements are much more difficult to meet at intermediate frequencies, let alone radio frequencies. Hence, digital integration of the second IF analog mixers (section 2.2.2) would require a more difficult and less optimal A/D converter in terms of performance and power consumption.

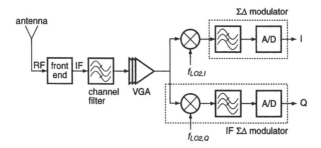

Fig. 2-7   *Receiver architecture with continuous-time baseband $\Sigma\Delta$ modulator (shown in upper path quadrature) and with continuous-time IF-to-baseband $\Sigma\Delta$ modulator (shown in lower quadrature path).*

In this book a technique is investigated to integrate the IF mixer with a lowpass $\Sigma\Delta$ modulator in the analog domain (chapter 4). This mixer and $\Sigma\Delta$ modulator topology will be referred to as IF-to-baseband $\Sigma\Delta$ modulator (or IF $\Sigma\Delta$ modulator). It is shown that the IF $\Sigma\Delta$ modulator (lower path Fig. 2-7) provides highly linear mixing, while power consumption is determined by the low-power ADC only. Moreover, two IF $\Sigma\Delta$ modulators in a quadrature configuration can be dynamically matched to improve image rejection performance (chapter 5). The dynamic element matching (DEM) algorithm is based on the complex bitstream output of the quadrature $\Sigma\Delta$ modulator (Fig. 2-8). The result is a low-power system with high image rejection.

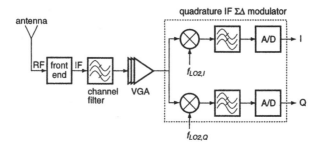

Fig. 2-8   *Quadrature IF-to-baseband $\Sigma\Delta$ modulator*

2.4   Performance parameters

In the previous section, the most important specifications for A/D converters embedded in a receiver architecture have been distinguished. These are dynamic range, linearity, image rejection, and power consumption. It was shown that the specifications strongly depend on the place of the A/D converter in the receiver. In this section, the ADC performance parameters are explained. In the final section, a figure-of-merit is defined, based on these parameters, for an objective comparison between different A/D converters.

2.4.1 Dynamic range

One of the important performance parameters of an A/D converter is dynamic range. In literature a number of different terms occur, indicating dynamic range performance, with only minor differences. The main definitions which are used throughout this book have been listed below. These are dynamic range, signal-to-noise ratio, peak SNR and resolution.

Dynamic range (DR) - ratio between maximum signal power and minimum detectable signal power within a specified bandwidth.

Signal-to-noise ratio (SNR) - ratio between signal power and noise power within a certain bandwidth.

Peak SNR - ratio between maximum signal power and noise power within a certain bandwidth.

Resolution - smallest output step, or least significant bit (LSB), indicating a change of the input signal.

It should be noted that these performance parameters are all relative numbers. Information about the (maximum) input power is needed for a complete qualification.

2.4.2 Linearity

A parameter which is particularly important in receivers is linearity. Non-linearity introduces distortion components, which should be weaker than the minimum detectable signal power. Different definitions, specifying non-linearity of the system, are distinguished. These are harmonic distortion, spurious-free dynamic range, intermodulation distortion, intermodulation intercept point, cross-modulation distortion and signal-to-noise-and-distortion ratio.

Harmonic distortion (HDx) - ratio between signal power and power of distortion component at x^{th} harmonic of signal frequency. Commonly, the second and third harmonic components, which are indicated as **HD2** and **HD3** respectively, are the most important (Fig. 2-9a).

Spurious-free dynamic range (SFDR) - ratio between maximum signal carrier and the maximum (in-band) distortion component (Fig. 2-9b).

Intermodulation distortion (IMx) - modulation of multiple signal tones, due to non-linearity, to spectral positions which are combinations of the signal frequencies. Fig. 2-9c shows a frequency spectrum with two signal tones at ω_1 and ω_2 respectively. Due to second-order and third-order distortion, intermodulation tones occur at spectral positions $\omega_2 - \omega_1$ (and $\omega_1 + \omega_2$) and $2\omega_1 - \omega_2$ (and $2\omega_2 - \omega_1$), respectively. Intermodulation distortion is defined as the distance between the carrier power of a signal tone and the power of the specified x^{th}-order intermodulation distortion tone. Again, the second-order and third-order intermodulation tones, which are defined by **IM2** and **IM3** respectively, are usually the most important.

Intermodulation intercept point (IPx) - (theoretical) signal carrier power for which the power of the specified x^{th}-order intermodulation distortion tone is equal to the signal carrier power (or in other words, the signal power for which IMx is 0dB). Again, **IP2** and **IP3** are the most important.

Cross-modulation distortion (CM) - modulation of spectral content around the carrier of an interference channel (right side of Fig. 2-9d), due to non-linearity, in the desired signal band (left side of Fig. 2-9d). Cross-modulation distortion is indicated as the distance between the signal carrier power and the cross-modulation distortion components.

Signal-to-noise-and-distortion ratio (SNDR) - ratio between signal power and total noise and distortion power within a certain bandwidth.

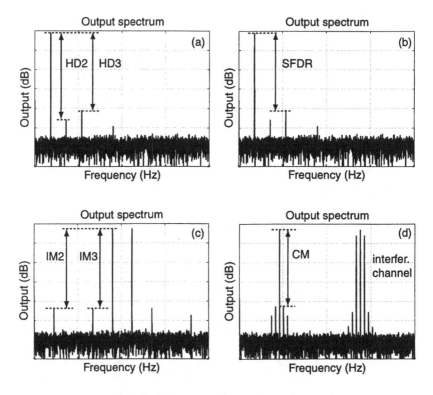

Fig. 2-9   *Harmonic distortion (a); Spurious-free dynamic range (b); Intermodulation distortion (c); Cross-modulation distortion (d).*

2.4.3 Image rejection

Frequency translation in a receiver, by means of a frequency multiplier or mixer, introduces the problem of image interference. Fig.2-10a shows the positive and negative frequency spectrum of two channels modulated around a frequency f_{LO2}. The light colored channel at the right of f_{LO2} is considered to be the desired signal, and the dark colored channel at the left of f_{LO2} is referred to as the image. When applied to a single mixer, operating at frequency f_{LO2}, both channels are upconverted and downconverted. The downconverted tones occur around DC. A single (or real) mixer is equally sensitive to signals at both sides of the LO carrier. In other words, the result of real frequency translation is a compound signal band, containing the spectral contents of both channels around f_{LO2} (Fig.2-10b). If the desired and image signals represent two different radio channels, a mixture of both channels would be audible after frequency translation. To handle this problem, which is known as image interference, the technique of quadrature mixing is used.

In the complex domain, a sinusoidal signal can be represented by the sum of a positive and a negative frequency component according to

$$\cos(\omega t + \varphi) = \frac{e^{j\omega t + \varphi}}{2} + \frac{e^{-j\omega t - \varphi}}{2}$$
$$\sin(\omega t + \varphi) = \frac{e^{j\omega t + \varphi}}{2j} - \frac{e^{-j\omega t - \varphi}}{2j} \tag{2-1}$$

where j is the complex operator, ω is the frequency (rad/s) and φ the phase (rad). It becomes clear from Eq. (2-1) that the positive and negative frequency terms of a cosine have equal phases, while the frequency terms of a sine are 180° out of phase. Moreover, comparing both signals shows that the phase differences between the positive and negative frequency terms of the cosine and sine signals are 270° and 90° respectively. This difference in phases is used to select either the positive or negative frequency component of a sinusoidal signal through complex adding or subtracting. This is shown mathematically by the following relations

$$\cos(\omega t + \varphi) + j\sin(\omega t + \varphi) = e^{j\omega t + \varphi}$$
$$\cos(\omega t + \varphi) - j\sin(\omega t + \varphi) = e^{-j\omega t - \varphi} \tag{2-2}$$

Consider the positive frequency term being the desired signal and the negative term as the image. By complex adding of quadrature signals, according to Eq. (2-2), the desired signal (positive frequency) is selected while the image (negative frequency) is cancelled out. Complete cancellation of the image occurs if the quadrature signals have

- *equal gains*
- *exactly 90° phase difference*

If the signals in Fig.2-10a are mixed by two identical mixers with 90° phase difference, quadrature signals are available at the output of the mixers. This implies that the desired signal can be distinguished from the image signal by complex adding of the quadrature mixer outputs. Fig.2-10c shows the complex frequency spectrum of the signals of Fig.2-10a after quadrature demodulation. In the positive frequency plane the desired signal band is present, while the image only appears in the negative frequency plane.

Due to mismatch, quadrature signals which are generated in the analog domain do not exactly meet the gain and phase requirements. Taking into account a relative gain error Δa and a phase error $\Delta\varphi$, Eq. (2-2) changes to

$$(1 + \Delta a)\cos(\omega t + \Delta\varphi) + j\sin(\omega t) \approx e^{j\omega t} + \frac{(\Delta a - j\Delta\varphi)}{2}e^{-j\omega t} \qquad (2\text{-}3)$$

for small error values. Eq. (2-3) shows that a suppressed image term is present. The distance between the power of the image in the negative plane and the 'leaked' image in the positive frequency plane is called the image rejection (*IR*)

$$IR \approx 10 \cdot \log\left(\frac{4}{\Delta a^2 + \Delta\varphi^2}\right) \qquad (2\text{-}4)$$

Example 2.1: In the case of a phase error of 1° and a gain error of 1% the image rejection is 40 dB, which means that the image signal is 40 dB lower than the carrier of the desired signal.

Fig.2-10d shows the complex spectrum of the signals of Fig.2-10a after quadrature frequency translation with a phase and gain error between the quadrature signals. The distance between the power of the image signal in the negative frequency plane, and the power of the image leakage in the positive frequency plane is the image rejection. For a state-of-the-art technology, typical image rejection ratios of about 45-50 dB can be achieved, based on matching. For many applications, this amount of image rejection is not sufficient. Other techniques can be used to further reduce image interference. The channel filter (Fig.2-2) already suppresses the image signal by 40 dB. Moreover, active or passive image rejection filters [2.18], [2.19], double quadrature mixers [2.20], [2.21], and dynamic element matching (chapter 5) effectively improve image rejection.

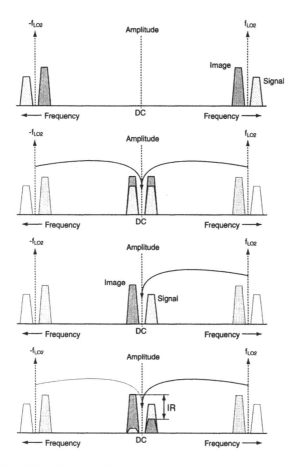

Fig. 2-10 *Double-sided frequency spectrum of two channels modulated around LO frequency f_{LO2} (a); after real frequency modulation with f_{LO2} (b); after ideal quadrature frequency modulation (c); after non-ideal quadrature frequency modulation (d).*

2.4.4 Figure-of-merit

With the parameters of the previous sections, the performance of different A/D converters can be compared. A useful tool for an objective comparison is the figure-of-merit (FOM). In literature, different FOMs can be found.

In [2.17], [2.22] an FOM is presented, expressing the power efficiency of an A/D converter in relation to the dynamic range (power) in a certain bandwidth

$$FOM = \frac{4kT \cdot DR \cdot BW}{P} \tag{2-5}$$

where T is temperature (K), k is Boltzman's constant (J/K), BW is the bandwidth (Hz) and P the power consumption (W) of the ADC. The FOM of Eq. (2-5) is equal to 1 for an integrator, implemented by an ideal class B amplifier [2.17]. This FOM does not contain linearity performance of the ADC, which is important for ADCs in receivers. Another FOM, which does include distortion, is related to the SNDR rather than DR

$$FOM = \frac{4kT \cdot SNDR \cdot BW}{P} \tag{2-6}$$

Because of the inclusion of distortion, the SNDR is always equal to or lower than the DR performance. Therefore, the FOM of Eq. (2-6) gives a lower value than the FOM of Eq. (2-5). With a FOM, only a rough comparison is made between different implementations. Other parameters, such as maximum input power, supply voltage or chip area, have not been taken into account and may be evenly important for particular applications. In chapter 6, different ADC designs from literature have been compared with the designs in this book, using the FOM of Eq. (2-6).

Example 2.2: Consider the following data: SNDR is 90 dB (10^9), BW is 100 kHz (10^5), and P is 0.1 W (10^{-1}). Substituting this data in Eq. (2-6) gives a FOM of $16.6*10^{-6}$.

2.5 GSM and AM/FM radio specifications

In Europe, the GSM transmit channels from a base cell station are modulated in the 935-960 MHz band, containing 125 channels of 200 kHz bandwidth. The modulation technique is Gaussian minimum-shift-keying (GMSK) with a data rate of 270.833 Kbits/sec. The minimum power of a desired channel at the antenna of the receiver is specified to be -104 dBm. To have sufficiently low bit error rate (BER), 9 dB of SNDR is required. This implies that the power of total in-band noise and distortion should be as low as -113 dBm. Total power of the adjacent interference channels in the GSM band can be up to -23 dBm, resulting in a dynamic range of 90 dB. Usually, the 200 kHz signal band is frequency translated to an IF of 10.7 MHz. Depending on the amount of AGC, typical DR and linearity requirements are 60-70 dB, after 200 kHz channel selection. Much better DR and linearity performance, in the order of 90 dB, is required, if A/D conversion is performed before analog channel selection and amplification.

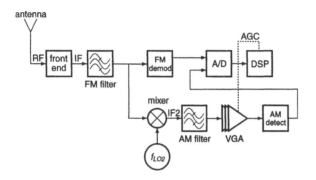

Fig. 2-11   *Model of AM/FM receiver with A/D converter*

The FM band, containing 100 kHz channels in Europe and 200 kHz channels in the United States, ranges from 65 MHz to 108 MHz. Fig.2-11 shows a block diagram of an AM/FM receiver with A/D converter. The FM channel is demodulated after channel filtering, typically at an IF of 10.7 MHz. Again, 10 to 11 bits of resolution in a 200 kHz FM band may be sufficient to achieve the noise requirements [2.23]. The AM receiver shares the front-end and channel filter of the FM receiver. Downconversion is performed to a second IF (IF2) of typically 455 kHz. Selectivity is obtained by a 9 kHz AM channel filter. The same ADC as for the FM digitizing can be used for AM digitizing. Shifting the ADC through the AM receive path yields the simplified system of Fig.2-12. AM and FM digitizing is performed at the 10.7 MHz IF and FM demodulation is done by the DSP. For FM reception, neighboring FM channels are suppressed by the FM channel filter. Depending on the amount of AGC, the DR at the input of the ADC is about 70 dB. For AM reception, neighboring AM channels are not suppressed by the FM channel filter. The filtered frequency band at the output of the 200 kHz FM channel filter contains about 20 AM channels. The required DR as well as the linearity of the ADC need to be larger than 90 dB to handle these AM channels.

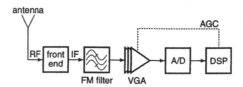

Fig. 2-12 *Model of FM/AM receiver with IF digitizing*

Some of the key issues in this book are dynamic range and linearity of the ADC. Increasing the performance of the ADC may result in relaxed requirements for the channel filter in the receiver path and a higher degree of integration. However, a cheaper channel filter with a low quality factor provides less filtering of direct neighboring channels. Therefore, suppression of the image channel, in the case of low-IF receivers, is also less effective. This implies that the image rejection requirement for the ADCs in a quadrature configuration becomes more stringent as well.

2.6   Summary

In this chapter some important performance parameters for the A/D converter in a radio receiver have been described. It was shown that the performance requirements strongly depend on the place of the A/D converter in the receiver. For reasons of cost, size and complexity reduction, an important trend in receiver design is digital integration of analog functionality, such as mixers and filters. Shifting the A/D converter towards the antenna side of the receiver results in tough requirements regarding dynamic range and linearity. These requirements are more difficult to meet at higher frequencies, due to circuit non-idealities and parasitic effects.

Baseband A/D converters in the traditional heterodyne receiver of Fig.2-2 have the most relaxed requirements. This is because interference signals have been filtered out by the channel filter and the desired channel is modulated down to DC or a low-IF frequency. At DC or a low-IF frequency, high linearity performance is achieved easily, as the low frequency (DC) gain of an amplifier can be high. Therefore, for high performance and low-power consumption of the A/D converter, the heterodyne receiver architecture of Fig.2-2 with the analog mixers should be preserved.

Digital integration of the channel filter in GSM or radio receivers puts heavy linearity (and image rejection) requirements on the VGA, quadrature mixers, anti-aliasing filters and A/D converters. A single-bit continuous-time baseband $\Sigma\Delta$ modulator has the ability of performing highly linear A/D conversion. Moreover, as sampling is done after filtering of the input signal by the continuous-time loopfilter, it has an inherent anti-aliasing filter. In this book, the combined design of a mixer and continuous-time $\Sigma\Delta$ modulator is investigated for high performance IF-to-baseband A/D conversion (chapter 4). With this A/D converter, channel filter requirements can be relaxed. Besides linearity and dynamic range, high image rejection performance is required as well. In chapter 5, the image rejection performance of the IF-to-baseband $\Sigma\Delta$ modulator in a quadrature configuration is investigated.

References

[2.1] Sevenhans, J., "The Single-Chip Digital Mobile Radio: Does it Really Make Sense?," *ISSCC Dig. Tech. Papers*, pp. 122-123, Feb. 1999.

[2.2] Kianush, K., C. Vaucher, "A global car radio IC with inaudible signal quality checks," *ISSCC Dig. Tech. Papers*, pp. 130-131, Feb. 1998.

[2.3] Breems, L.J., E.J. van der Zwan and J.H. Huijsing, "A 1.8 mW CMOS ΣΔ Modulator for Mobile Communication," *Proc. of Workshop on Circuits, Systems and Signal Processing*, pp. 61-64, Nov. 1998.

[2.4] Liu, Chia-Liang, "Impacts of I/Q imbalance on QPSK-OFDM-QAM detection," *Proc. of Int. Conf. on Consumer Electronics*, pp. 384-385, July, 1998.

[2.5] Jantzi, S.A., K.W. Martin, and A.S. Sedra, "Quadrature Bandpass ΔΣ Modulation for Digital Radio," *IEEE J. Solid-State Circuits*, vol. 32, pp. 1935-1950, Dec. 1997.

[2.6] Singor, F.W., W.M. Snelgrove, "Switched-Capacitor Bandpass Delta-Sigma A/D Modulation at 10.7 MHz," *IEEE J. Solid-State Circuits*, vol. 30, pp. 184-192, March 1995.

[2.7] Song, B.S., "A Fourth-Order Bandpass Delta-Sigma Modulator with Reduced Number of Op Amps," *IEEE J. Solid-State Circuits*, vol. 30, pp. 1309-1315, Dec. 1995.

[2.8] Ong A.K., B.A. Wooley, "A Two-Path Bandpass ΣΔ Modulator for Digital IF Extraction at 20 MHz," *IEEE J. Solid-State Circuits*, vol. 32, pp. 1920-1934, Dec. 1997.

[2.9] Hairapetian, A., "An 81 MHz IF Receiver in CMOS," *ISSCC Dig. Tech. Papers*, pp. 56-57, Feb. 1996.

[2.10] Vorenkamp, P., R. Roovers, "A 12-b, 60-MSample/s Cascaded Folding and Interpolating ADC," *IEEE J. Solid-State Circuits*, vol. 32, pp. 1876-1886, Dec. 1997.

[2.11] Kwak, S-U, B-S. Song, K. Bacrania, "A 15-b, 5-Msample/s Low-Spurious CMOS ADC," *IEEE J. of Solid-State Circuits*, vol. 32, pp. 1866-1875, Dec. 1997.

[2.12] Gao, W., W.M. Snelgrove, "A 950-MHz IF Second-Order Integrated LC Bandpass Delta-Sigma Modulator," *IEEE J. Solid-State Circuits*, vol. 33, pp. 723-732, May 1998.

[2.13] Cherry, J.A, W.M. Snelgrove, *Continuous-time delta-sigma modulators for high-speed A/D conversion; theory, practice and fundamental performance limits*, Kluwer Academic Publishers, Boston, 2000.

[2.14] Fujimori, I., K. Koyama, D. Trager, F. Tam, L. Longo, "A 5-V Single-Chip Delta-Sigma Audio A/D Converter with 111 dB Dynamic Range," *IEEE J. Solid-State Circuits*, vol. 32, pp. 329-336, March 1997.

[2.15] Ritoniemi, T., E. Pajarre, S. Ingalsuo, T. Husu, V. Eerola, T. Saramaki, "A Stereo Audio Sigma-Delta A/D-Converter," *IEEE J. Solid-State Circuits*, vol. 29, pp. 1514-1523, Dec. 1994.

[2.16] Coban, A.L., P.E. Allen, "A 1.5 V 1.0 mW Audio $\Delta\Sigma$ Modulator with 98 dB Dynamic Range," *ISSCC Dig. Tech. Papers*, pp. 50-51, Feb. 1999.

[2.17] Rabii, S., B. Wooley, "A 1.8-V digital-audio Sigma-Delta modulator in 0.8µm CMOS," *IEEE J. Solid-State Circuits*, vol. 32, pp. 783-796, June 1997.

[2.18] Voorman, J.O., "Continuous-time analog integrated filters," in *Integrated Continuous-Time Filters*, IEEE Press, New York, 1993.

[2.19] Gingell, M.J., "Single sideband modulation using sequence asymmetric polyphase networks," *Elect. Commun.*, vol. 48, pp. 21-25, 1973.

[2.20] Crols, J. M.S.J. Steyaert, "A Single-Chip 900 MHz CMOS Receiver Front-End with a High Performance Low-IF Topology," *IEEE J. Solid-State Circuits*, vol. 30, pp. 1483-1492, Dec. 1995.

[2.21] Rudell, J.C., J.J. Ou, T.B. Cho, G. Chien, F. Brianti, J. A. Weldon, P.R. Gray, "A 1.9-GHz Wide-Band IF Double Conversion CMOS Receiver for Cordless Telephone Applications," *IEEE J. Solid-State Circuits*, vol. 32, pp. 2071-2088, Dec. 1997.

[2.22] Sansen, W., Advanced engineering course on low-voltage low-power analog CMOS design, Lausanne, June 21-25, 1999.

[2.23] Zwan, E.J. van der, K. Philips, C. Bastiaanse, "A 10.7 MHz IF-to-Baseband $\Sigma\Delta$ A/D Conversion System for AM/FM Radio Receivers," *ISSCC Dig. Tech. Papers*, pp. 340-341, Feb. 2000.

Continuous-time sigma-delta modulation

3

3.1 Introduction

In 1954, a patent was filed by Cutler [3.1] of a feedback system with a low-resolution quantizer in the forward path. The quantization error of a quantizer was fed back and subtracted from the input signal. This principle of improving the resolution of a coarse quantizer by use of feedback is the basic concept of a delta-sigma or sigma-delta converter [3.2]. In 1962, Inose et al. [3.3] came up with the idea of adding a filter in the forward path of a delta modulator [3.4], in front of the quantizer. This system was called a 'delta-sigma modulator', where 'delta' refers to the delta-modulator, and 'sigma' refers to summation by the integrator.

In this chapter, the theory and design of a continuous-time sigma-delta modulator is explored. In section 3.2 the basic theory of sigma-delta modulation is explained. Stability analysis of a linear model of a continuous-time sigma-delta modulator is performed in section 3.3 and in section 3.4 the filter design of higher-order lowpass filters has been elaborated. Finally, the design of a complex filter for a quadrature sigma-delta modulator architecture is presented in section 3.5.

3.2 Theory of sigma-delta modulation

In this section, the basic theory of $\Sigma\Delta$ modulation is described and the main non-idealities which determine the overall performance. The principles of oversampling and noise-shaping are explained in section 3.2.1. The error of the quantization process is assumed to be white noise for simplicity of the analysis [3.5]. However, analyzing the quantization spectrum (section 3.2.2) shows that it is far from white, but contains periodic, offset-related sequences which introduce tones in the output spectrum. As noise imposes a limit on the minimum detectable input signal, distortion affects the performance for the maximum input. In section 3.2.3, the effect of distortion due to non-linearity of the input stage of the loopfilter is calculated. Intersymbol interference (section 3.2.4) is caused by waveform asymmetry of the feedback pulses and forms another source of non-linearity in continuous-time $\Sigma\Delta$ modulators. Besides quantization noise and tones, phase jitter of the reference clock also adds noise to the system, which decreases the maximum obtainable signal-to-noise ratio (section 3.2.5). Finally, the subject of aliasing is discussed in section 3.2.6, as it is of great importance for a $\Sigma\Delta$ modulator when used in a receiver (section 2.2.1).

3.2.1 Oversampling and noise-shaping

The principle of operation of a $\Sigma\Delta$ modulator is explained by the block diagram of a single-loop $\Sigma\Delta$ modulator (Fig. 3-1). The modulator consists of a loopfilter which in its simplest form is an accumulator or integrator. The filter output is sampled and quantized by an A/D converter which introduces a quantization error E. The digital output signal is subtracted from the analog input via a D/A converter in the feedback path. The error E due to the quantization process is the difference between the analog quantizer input signal and the quantized output. Suppression of the quantization error in a $\Sigma\Delta$ modulator is provided by two mechanisms: oversampling of the signal bandwidth and shaping of the noise by the loopfilter. Because the reduction of the quantization error by these mechanisms is so effective, a high-resolution digital output is obtained, using only a low-resolution quantizer. In the extreme case a one-bit quantizer with two output levels is sufficient. In this section, the principles of oversampling and noise-shaping are briefly introduced.

Oversampling

A one-bit quantizer generates a bitstream pattern with output levels $\pm q/2$, where q is the quantization step size. The bitstream spectrum contains information about the input signal and a quantization error, which is introduced by the quantizer.

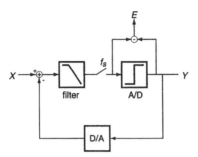

Fig. 3-1   *Model of a single-loop continuous-time ΣΔ modulator*

For simplicity, the quantization error is assumed to have a white noise spectrum [3.5] and the total power of the quantization error signal equals [3.2]

$$e^2_{rms} = \frac{q^2}{12} \tag{3-1}$$

The power spectral density of the sampled quantization error signal is

$$E(f) = \frac{q^2}{6f_s}, 0 \leq f < \frac{f_s}{2} \tag{3-2}$$

where f_s is the sampling frequency. This equation shows the relation between the quantization noise power and sampling frequency. It shows that the higher the sampling frequency, the lower the noise power density. This is the principle of oversampling. The integrated noise power within a fixed signal bandwidth decreases if the sampling rate is increased. In mathematical terms, the total in-band quantization noise power equals

$$N_q = \int_0^{f_b} E(f)df = \frac{e^2_{rm}}{m} \tag{3-3}$$

where f_b is the signal bandwidth. The oversampling ratio m is the ratio between the sampling frequency and the Nyquist bandwidth (twice the signal bandwidth)

$$m = \frac{f_s}{2f_b} \tag{3-4}$$

The dynamic range is defined as the ratio between the in-band noise power and the maximum power of the input signal. The maximum input power is about a factor of 4 smaller than the quantizer output power [3.6]

$$P_{in,\,max} = \frac{(q/2)^2}{4} \tag{3-5}$$

Noise-shaping

For low frequency signals, the DAC in the feedback path has a gain of approximately 1. Fig.3-2 shows a highly simplified linear model of a continuous-time $\Sigma\Delta$ modulator. Using this model, the output $Y(s)$ of the $\Sigma\Delta$ modulator is determined

$$Y(s) = \frac{H(s)}{1 + H(s)} \cdot X(s) + \frac{1}{1 + H(s)} \cdot E(s) \tag{3-6}$$

where $X(s)$ is the analog input signal and $H(s)$ is the loopfilter transfer function. The first term of the right-hand side of Eq. (3-6) is the signal transfer function (STF) and the second term is the noise transfer function (NTF). If $H(s)$ has a lowpass filter characteristic with high DC gain, then for low-frequencies the signal transfer function is close to 1, while the quantization error tends toward zero (NTF is 0). For frequencies close to half the sampling frequency, the input signal is filtered and the quantization error becomes large. This shows that the quantization noise density function is not constant over frequency, but has a shaped frequency spectrum. This is the principle of noise-shaping.

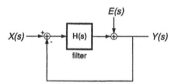

Fig. 3-2   *Simplified linear continuous-time $\Sigma\Delta$ modulator model*

In the next section, the quantization error spectrum is described in more detail. Although, the white noise assumption is useful for estimating the in-band quantization error power, it is shown that the quantization error signal consists of signal dependent sequences, with periodic, offset-related tones for small input signals.

3.2.2 Tones

Tones are spectral components that are induced by DC offsets, which generate periodic quantization error sequences. These periodic quantization errors occur as discrete peaks in the frequency spectrum with input signal dependent frequencies. The frequencies of tones are calculated by the empirical relation [3.2]

$$f_n = nf_s \cdot \frac{|V_{os}|}{2\Delta_q}, n = \{0, 1, 2, ...\} \tag{3-7}$$

where n is an integer, f_s is the sampling frequency, V_{os} is the DC offset voltage and Δ_q is the quantization step size. This phenomenon has been simulated with a model of a fourth-order 1-bit sigma-delta modulator. Fig. 3-3 shows the simulated time domain quantization error signal in the case of an offset of 3 mV (Δ_q is 1 V) and a 12.96 MHz sampling frequency. Fig. 3-3 reveals that the quantization error is clearly not a random signal but, on the other hand, that periodic sequences are not easily recognized. The autocorrelation technique offers more insight into the presence of tones in the time domain signal [3.2]. With (discrete-time) autocorrelation, the interdependence between a fixed quantization error sequence and a sequence shifted in time by l sample periods is determined as a function of l.

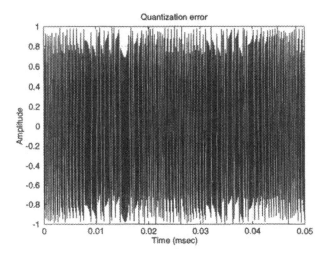

Fig. 3-3   *Quantization error signal of fourth-order 1-bit $\Sigma\Delta$ modulator for 3 mV DC input (Δ_q is 1 V)*

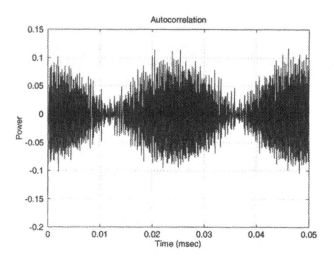

Fig.3-4 *Autocorrelation of quantization error signal of Fig. 3-3*

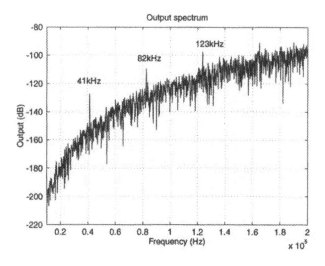

Fig. 3-5   *Output spectrum of fourth-order 1-bit $\Sigma\Delta$ modulator for 3 mV DC input (Δ_q is 1 V)*

Fig. 3-4 shows the autocorrelation of the quantization error waveform of Fig. 3-3. A periodic sequence can be observed with a period time of 24.5 μs (41 kHz). Fig. 3-5 shows the low frequency output spectrum of the ΣΔ modulator with the DC input. A tone of -128 dB occurs at 41 kHz and the harmonics are at 82 kHz, 123 kHz, and 164 kHz (n is 4, 8, 12 and 16). Moreover, Fig. 3-5 shows that the energy of the tones increases with the frequency. This is due to the decrease in loopgain at higher frequencies. Therefore, the strongest tone occurs at the end of the signal band. This can be a serious problem if the gain is not sufficient.

A method to manipulate tones is the use of zeros in the NTF characteristic that comply with the inverse Chebyshev placement [3.2]. This way, the NTF gain is distributed equally within the signal band of interest. These inverse Chebyshev zeros can be made by adding local feedback (resonator) paths in the loopfilter. This will be treated in detail in section 3.4.2. Another way of manipulating tones is applying a dither signal [3.7] to the input, to decorrelate the quantization errors. This is beyond the scope of this book.

So far, only baseband tones have been investigated. However, at high frequencies (near half the sampling frequency) tones are present as well and are very strong due to the lack of loopgain at those frequencies. The tones are situated at

$$f_n = \frac{f_s}{2}\left(1 - n\frac{|V_{os}|}{\Delta_q}\right), n = \{1, 2, 3, \ldots\} \tag{3-8}$$

The high frequency tones can be demodulated into baseband by even-order harmonic distortion [3.6] or parasitic coupling of a clock signal with a frequency near half the sampling rate [3.8], [3.9]. In the case of second harmonic distortion the baseband tones occur at

$$f_{tones} = |f_{n1} - f_{n2}|, |f_{n2} - f_{n1}| \tag{3-9}$$

where f_{n1} and f_{n2} are two high frequency tones. In the case of modulation with half the sampling frequency the in-band tone will be at

$$f_{tone} = \left|\frac{f_s}{2} - f_n\right| \tag{3-10}$$

Fig. 3-6 shows the simulated quantization spectrum near half the sampling frequency for a 3mV DC input. A large tone of -10 dB occurs at 6.4795 MHz, which is located at an offset frequency of 20.5 kHz away from half the sampling frequency.

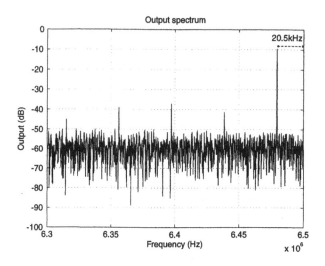

Fig. 3-6   *High frequency output spectrum near $f_s/2$ of $\Sigma\Delta$ modulator with 3 mV DC input*

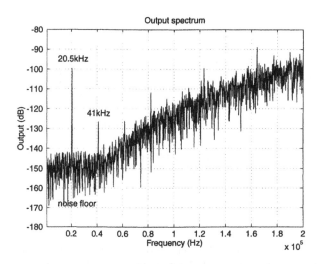

Fig. 3-7   *Output spectrum of $\Sigma\Delta$ modulator for 3 mV DC input and 0.003% modulation at $f_s/2$*

When this tone is modulated with half the sampling frequency, it will be downconverted to 20.5 kHz. In fact, a fraction of the entire noise band near half the sample frequency will 'leak' into baseband. This can be seen from the simulated spectrum of Fig.3-7. This plot shows the low frequency band of the output spectrum if a fraction (0.003%) of the bitstream is modulated with half the sampling frequency. A -100 dB tone occurs in the low-frequency spectrum at 20.5 kHz. Also, the low-frequency noise floor has been raised considerably, compared to the spectrum of Fig.3-5. This is an important effect which deteriorates performance easily and careful design attention is demanded. The subject of high frequency quantization noise demodulation is discussed in more detail in chapter 4.

3.2.3 Harmonic distortion

In the previous sections, various sources of noise which determine the theoretical limit of the signal-to-noise ratio have been highlighted. Another performance parameter is the SNDR, which is particularly important for radio applications as has been explained in chapter 2. In a single-bit modulator the DAC in the feedback loop is perfectly linear, as it switches between two quantization levels only. Therefore, harmonic distortion mainly occurs as a result of non-linearity of the active elements in the loopfilter. Furthermore, linearity requirements for the input stage of the loopfilter are most stringent, as distortion of the higher order stages is suppressed by the gain of the previous stages. In this section, a relation between the distortion parameter of the input loopfilter stage and the distortion of the $\Sigma\Delta$ modulator is derived [3.10].

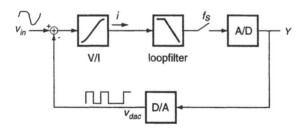

Fig. 3-8   *Single-bit $\Sigma\Delta$ modulator with non-linear input stage*

Basically, a continuous-time integrator consists of a capacitor which is charged or discharged by a current signal. Considering voltage input (v_{in}) and voltage feedback signals (v_{dac}), a voltage-to-current (V/I) converter is needed to drive the first stage of the loopfilter (Fig. 3-8). The active V/I converter is a non-linear device and the loopfilter input current i as a function of the input voltage v_{in} and the DAC voltage v_{dac} can be expressed as

$$i = \sum_{n=1}^{\infty} g_n \cdot (v_{in} - v_{dac})^n \qquad (3\text{-}11)$$

where g_n is the nth-harmonic transfer coefficient of the V/I converter. Furthermore, the modulator is supposed to be differential (no even order harmonics) and if the 3rd harmonic is the dominant non-linearity, Eq. (3-11) simplifies into

$$i \approx g_1 \cdot (v_{in} - v_{dac}) + g_3 \cdot (v_{in} - v_{dac})^3 \qquad (3\text{-}12)$$

The error voltage $v_{in} - v_{dac}$ is the difference between a sinusoidal (input) signal and a rectangular (DAC) signal (in the case of a single-bit DAC). In mathematical terms the error signal is

$$v_{in} - v_{dac} = \hat{V}_{in} \cdot \cos(\omega_i t) - \pm 1 \cdot V_{dac} \qquad (3\text{-}13)$$

where V_{in} is the input amplitude, ω_i is the input signal frequency and V_{dac} is the reference level of the DAC. Keeping in mind that the second harmonic distortion term of the DAC signal is a positive constant and that the third harmonic is equal to the ground harmonic, the third harmonic of Eq. (3-13) equals

$$(v_{in} - v_{dac})^3 \approx \frac{\hat{V}_{in}^3}{4} \cos(3\omega_i t) - (\pm 1 \cdot V_{dac}) \cdot \frac{3\hat{V}_{in}^2}{2} \cos(2\omega_i t) \qquad (3\text{-}14)$$

discarding the ground harmonic terms. Due to the oversampling and the high gain, a sigma-delta modulator closely tracks low-frequency input signals. Therefore, for frequencies much lower than the sampling frequency the following relation holds

$$\pm 1 \cdot V_{dac} \approx \hat{V}_{in} \cos(\omega_i t), \ \omega_i \ll \omega_s \qquad (3\text{-}15)$$

Substituting this result into Eq. (3-14) yields an expression for the third-harmonic of the input signal

$$\left|(v_{in} - v_{dac})^3\right| \approx \frac{\hat{V}_{in}^{\,3}}{2} \cos(3\omega_i t) \tag{3-16}$$

Combining Eq. (3-12), Eq. (3-13) and Eq. (3-16) yields the relation between the output current and input voltage of a non-linear V/I converter in a $\Sigma\Delta$ modulator. The third harmonic component can be referred to the input of the V/I converter by dividing the second right-hand term of Eq. (3-12) by the linear transfer coefficient g_1.

$$v_{3^{rd}} \approx \frac{g_3}{g_1} \cdot \frac{\hat{V}_{in}^{\,3}}{2} \cos(3\omega_i t) \tag{3-17}$$

Due to the high gain at low frequencies and the large amount of oversampling, the $\Sigma\Delta$ loop also compensates low-frequency harmonic components. Consequently, an extra signal at the third harmonic frequency appears in the output spectrum that is equal to Eq. (3-17). The (low-frequency) third harmonic distortion (HD3) of the continuous-time $\Sigma\Delta$ modulator of Fig. 3-8 equals

$$HD3 \approx \frac{g_3}{g_1} \cdot \frac{\hat{v}_{in}^{\,2}}{2} \tag{3-18}$$

Example 3.1: To verify Eq. (3-18), the system of Fig. 3-8 has been simulated where g_1 is 1, g_3 is 0.0001, and V_{in} is 0.707 (calculated $HD3$ is -92 dB). Fig. 3-9 shows the simulated bitstream spectrum. The third-harmonic component (30kHz) is 92 dB down relative to the signal carrier (10 kHz) which is in agreement with the calculated result of Eq. (3-18).

It should be noted that this analysis is only valid for low-frequency signals in the band where the gain of the loopfilter is high. Fig. 3-10 shows the same simulation as in the example, only with a 33.33 kHz input signal. The third-harmonic is at 100 kHz where loopfilter gain drops with the order of the loopfilter (4 in this case). Because the loopfilter gain drops, the tone at the third-harmonic frequency of the 33.33 kHz signal component is less suppressed. This effect is not related to the non-linearity of the integrator and also occurs in an ideal distortion-free $\Sigma\Delta$ modulator.

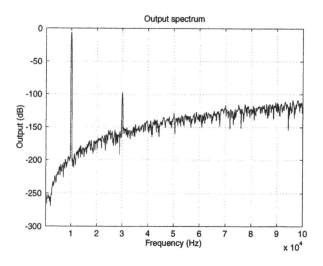

Fig. 3-9 *Output spectrum of ΣΔ modulator with non-linear input stage (0.01%) and 10 kHz input signal*

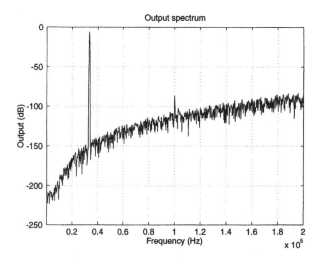

Fig. 3-10 *Output spectrum of ΣΔ modulator with non-linear input stage (0.01%) and 33.33 kHz input signal*

This type of non-linearity is reduced by using a loopfilter with local feedback paths (section 3.4.2). With this filter topology an inverse Chebyshev characteristic can be designed rather than a Butterworth filter. The resonator gains are evenly distributed in the signal band to reduce the quantization noise (and distortion components) at the end of the signal band and to have a maximally flat quantization noise floor in the signal band.

Another method to suppress the distortion of the non-linear input V/I converters is to use a multi-bit $\Sigma\Delta$ modulator. With a multi-bit DAC, the feedback and input signals cancel better than in case of a single-bit feedback signal. However, due to mismatch between the quantization levels, a multi-bit DAC is non-linear and introduces distortion itself.

3.2.4 Intersymbol interference

In section 3.2.2 it was explained that high frequency tones in the quantization error spectrum fold back into baseband by even-order harmonic distortion. Even-order harmonic distortion occurs as a result of imbalances in a non-linear differential system. A well-known type of imbalance in a continuous-time $\Sigma\Delta$ modulator is asymmetry between the positive and negative DAC feedback pulses, also referred to as intersymbol interference [3.11]. This is due to different transition times when switching from the positive to the negative reference and vice versa (Fig. 3-11a). Assuming that the rising edge is steeper than the falling edge, the charge transfers during transition intervals τ_f and τ_r do not completely balance out. The result of waveform asymmetry is that the energy contents of e.g. a 0101 pattern and a 1100 pattern are different. This signal dependent imbalance causes an offset and even-order harmonic distortion.

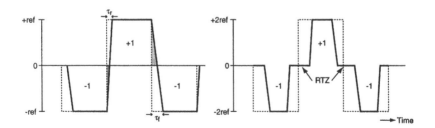

Fig. 3-11 *Feedback signal with waveform asymmetry (solid line), due to different rise and fall times τ_r and τ_f (a); Feedback signal with return-to-zero (b).*

Example 3.2: Fig.3-12 shows the simulated result of a $\Sigma\Delta$ modulator with a relative waveform asymmetry error of 0.1% of a full DAC pulse. The input signal is at 10 kHz. The second harmonic tone at 20 kHz is 75 dB down relative to the signal carrier. The noise floor is raised to -120 dB compared to the spectrum of Fig.3-5 because of demodulation of high frequency quantization noise.

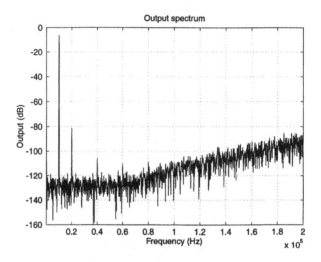

Fig. 3-12 *Simulated output spectrum of $\Sigma\Delta$ modulator with 0.1% waveform asymmetry*

Waveform asymmetry can be reduced by short transition intervals and matching of the transition edges. However, this may not be easy to realize, especially at higher sampling frequencies. A more robust method of improving matching of waveforms, is the return-to-zero (RTZ) technique. This technique is shown in Fig.3-11b and is based on the principle of resetting each feedback pulse for a fraction of the sampling period. As a result, all pulses have a positive and a negative edge, which greatly reduces signal dependence of the mismatch. Now, any mismatch between the positive and negative pulses only affect the offset. If the RTZ interval has a duty-cycle δ_{rtz}, the reference level of the feedback pulse should increase with the inverse factor in order to have the same charge. Fig.3-11b shows a feedback pulse with a 50% duty cycle ($\delta_{rtz} = 0.5$).

3.2.5 Phase jitter

Fig. 3-11b shows the waveform of a typical feedback signal, with a periodic return-to-zero state, in a single-bit continuous-time $\Sigma\Delta$ modulator. In the case of zero rise and fall times, the charge of the feedback pulse is determined by the amplitude of the DAC current I_{dac} and the pulse width δT_s, where δ is the pulse duty-cycle, with a value between 0 and 1

$$Q = I_{dac} \cdot \delta T_s \qquad (3\text{-}19)$$

Any inaccuracy in the reference charge has a direct impact on the performance of the $\Sigma\Delta$ modulator. As described in the previous section, waveform asymmetry is a cause for in-band distortion and noise. Besides the inaccuracy of the waveform, also timing uncertainty affects the reference charge in Eq. (3-19). Timing uncertainty is caused by phase noise in the oscillator. A commonly used measure of the total amount of timing noise in the frequency range of interest is the jitter specification [3.12]. Assuming that jitter can be considered as white noise, with standard deviation σ_j, the jitter contributed charge Q_j can be expressed by [3.6]

$$Q_j = \sigma_j I_{dac} \qquad (3\text{-}20)$$

To determine the noise power in the signal band, the power spectral density of the jitter is calculated. The noise power in the signal band can be calculated by

$$N_j^2 = \int_{-f_b}^{f_b} \frac{Q_j^2}{f_s} df = (\sigma_j I_{dac})^2 \cdot \frac{2f_b}{f_s} \qquad (3\text{-}21)$$

The SNR due to clock jitter is calculated by the ratio between the maximum input signal power (Eq. (3-5)) and the jitter power

$$SNR = 10\log\left(\frac{\delta^2}{4} \cdot \frac{m}{\sigma_j^2 \cdot f_s^2}\right) \qquad (3\text{-}22)$$

Example 3.3: In the case of a 50% duty cycle ($\delta = 0.5$), an oversampling ratio m of 64 and $\sigma_j = 0.05\%$ of the sampling frequency, the calculated SNR is 72 dB. Fig. 3-13 shows the simulated output spectrum of a $\Sigma\Delta$ modulator with clock jitter. The quantization noise is not dominant in this simulation. The simulated SNR equals 71.8 dB which agrees with the theoretical result (white noise assumption).

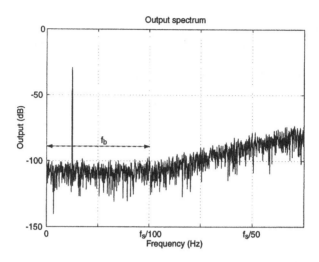

Fig. 3-13 *Simulated output spectrum of ΣΔ modulator with phase jitter*

It should be noted that Eq. (3-22) only applies to a feedback pulse with a periodic return-to-zero state. In the case of a full period feedback pulse without return-to-zero, the jitter power depends on the relative number of edges in a bitstream sequence. This can be easily understood considering a 11110000 and a 01010101 bitstream sequence (no return-to-zero). A timing uncertainty only occurs at the edge when the feedback pulse changes from 0 to 1 or vice versa. Obviously, the jitter power of the first sequence is smaller than the second one, due to the difference in number of transitions (1 versus 7). This implies that the jitter power is signal dependent and becomes larger for smaller input signals.

3.2.6 Aliasing

Due to the sampling process, any input frequency ω_i which is larger than half the sampling frequency ω_s will reflect into the frequency range $0 < \omega < \omega_s/2$. This is the known problem of aliasing. Reduction of the aliasing effect is achieved by means of an analog filter in front of the sampler, to suppress the energy content of the frequency bands at multiples of the sampling frequency. An important property of a continuous-time ΣΔ modulator is that the sampling takes place at the output of the continuous-time loopfilter (in contrast to a discrete-time ΣΔ modulator where sampling is done at the input of the loopfilter). The loopfilter operates as an anti-aliasing filter, therefore discarding the need for an anti-aliasing filter in front of the ΣΔ modulator. The suppression of an aliasing component, originating from an input signal with a frequency close to the sampling frequency, can be easily calculated, using the linear model of a continuous-time ΣΔ modulator (Fig. 3-15). Under the assumption that the hold function in the feedback path is a lowpass filter with a gain of 1 at low frequencies and a gain of 0 for frequencies near the sampling frequency, it is shown in [3.6] that the aliasing component of an input signal with frequency $f_s - f_b$ is reduced by the loopfilter with a factor

$$\frac{H(f_b)}{H(f_s - f_b)} \tag{3-23}$$

where $H(f_b)$ is the loopfilter gain for the low frequency aliased component at f_b and $H(f_s\text{-}f_b)$ the gain for the input frequency close to the sampling frequency.

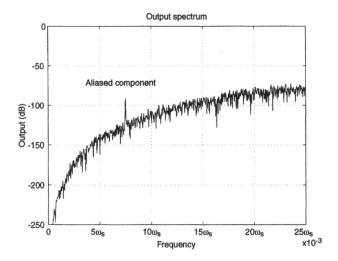

Fig. 3-14 *Simulated   low-frequency   output   spectrum   of   a continuous-time ΣΔ modulator with -9 dB input signal at ω_s-0.0075ω_s*

Example 3.4: A ΣΔ modulator with the fifth-order feedforward compensated loopfilter characteristic of Fig. 3-24 has been simulated. The simulation result is shown in Fig. 3-14 for a -9 dB input signal at ω_s-0.0075ω_s. The aliased component occurs at 0.0075ω_s, and is suppressed by about 80 dB.

3.3 Linear stability analysis

Stability is an important issue in the design of a $\Sigma\Delta$ modulator and yet, due to the fact that the quantizer is a non-linear element, it has not been possible to find a mathematical solution for guaranteed loop stability of higher-order modulators. However, a useful insight to stability is obtained by the analysis of limit cycles, or idling tones, in the output signal of the $\Sigma\Delta$ modulator. A stable $\Sigma\Delta$ modulator exhibits out-of-the-band limit cycles at fractions of the sampling frequency which do not affect the SNR [3.13]. In contrast, the SNR of an unstable modulator is deteriorated by low-frequency limit cycles, due to an output of alternating long strings of 1s and 0s [3.2] that may have unbounded states. Numerous methods for stability analysis and rules of thumb have been presented [3.14]- [3.15] which are valuable for higher-order $\Sigma\Delta$ modulator design. In this section a linear quantizer model is used for stability analysis [3.16], [3.17]. With this linear model the root locus method can be applied to analyze the linearized system for stable limit cycles. First, stability of a $\Sigma\Delta$ modulator in the case of a small input signal is discussed, followed by the large signal stability analysis.

3.3.1 Small signal stability

Fig.3-15 shows a linear model of a continuous-time (CT) $\Sigma\Delta$ modulator. The system consists of a CT loopfilter with transfer function $H(s)$ and a sampled quantizer. The quantizer is modeled with a linear gain κ and a phase shift θ [3.18]. The quantizer output, which is a train of pulses, is fed back through a D/A converter, being a zero-order hold function, and subtracted from the analog input signal.

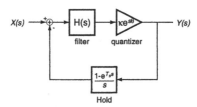

Fig. 3-15 *Linear model of a single-loop continuous-time $\Sigma\Delta$ modulator*

It is assumed that the input of the quantizer is a sinusoidal waveform. The gain κ of the quantizer can be defined as the ratio between the input power and the output power of the quantizer. The sampling of the quantizer introduces a frequency dependent phase uncertainty in the loop [3.18] for input frequencies which are rational fractions of the sampling frequency [3.19]. This means that for a certain range of phases, the quantizer produces the same output pattern, regardless of the phase of the input signal. This is explained by Fig.3-16 for a sinusoidal quantizer input signal with frequency $f_s/4$ in the case of a one-bit quantizer. Consider the sinusoidal signal in Fig.3-16 (solid line). Both at sample moments nT_s and $(n+1)T_s$, the quantizer gives a positive output. At sample moment $(n-1)T_s$ the quantizer output is negative. After four sample moments the output pattern is repetitive. If the (solid) sinusoidal signal is shifted in phase over a range of $\pi/4$ (dotted lines in Fig.3-16) the quantizer output gives the exact same output pattern for the whole range of phases. Thus, the quantizer is not able to detect the phase of an input signal at $f_s/4$ and therefore has a phase uncertainty. At $f_s/4$ the maximum phase uncertainty is $\pi/4$. This uncertainty in phase of the quantizer is largest at $f_s/2$ ($\pi/2$). In the Laplace domain, the quantizer is modeled as

$$H_q(s) = \kappa e^{s\theta} \qquad (3\text{-}24)$$

where κ is the quantizer gain and θ the phase shift.

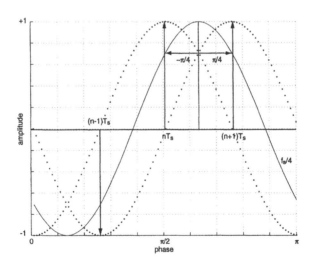

Fig. 3-16 *Maximum phase shift for sine wave at $f_s/4$ without changing the sampled output*

For signals that are not rational fractions of the sampling frequency, any shift in phase always gives a change in the quantizer output pattern. Therefore, the phase uncertainty for those frequencies is zero.

The DAC in the feedback path has the function of a zero-order hold

$$H_{DAC}(s) = \frac{1 - e^{-sT_s}}{s} \tag{3-25}$$

With the linear model of Fig. 3-15, the $\Sigma\Delta$ modulator can be analyzed using the root locus method, which shows the closed loop poles as a function of the quantizer gain for different phase shifts of the quantizer. The modulator is stable if all poles are in the left-half plane of the root locus plot. Stability of a $\Sigma\Delta$ modulator becomes an issue if the loopfilter order is larger than 1. In the following, stability of a $\Sigma\Delta$ modulator with a second-order loopfilter is discussed. A second-order lowpass loopfilter consists of two integrator stages which are connected in cascade. The transfer function of a second-order integrator filter is given by

$$H_f(s) = \frac{1}{s^2} \tag{3-26}$$

The filter characteristic has two poles and for simplicity, the DC gains of both integrators have been considered to be infinite (ideal integrators). With the transfer functions of Eq. (3-24), Eq. (3-25), and Eq. (3-26), the characteristic equation of the linear system of Fig. 3-15 can be determined

$$1 + 2\kappa e^{s\theta}\left(\frac{1 - e^{-sT_s}}{s}\right) \cdot \left(\frac{1}{s^2}\right) = 0 \tag{3-27}$$

Due to the exponential terms in Eq. (3-27) there is no analytical expression which can solve this characteristic equation. With numerical analysis, the root locus of the second-order system has been determined (Fig. 3-17). Fig. 3-17 shows that the poles which are starting from the origin, move directly into the right-half plane for any gain κ and any phase θ. Hence, the linear model of a second-order $\Sigma\Delta$ modulator is unstable for any quantizer gain and phase. To stabilize the second-order modulator, a zero can be introduced in the filter transfer function of

Eq. (3-26), by means of feedforward or feedback compensation (section 3.4). An example transfer function of such a second-order low-pass filter is given by

$$H_f(s) = \frac{2(s + 0.5)}{s^2} \qquad (3-28)$$

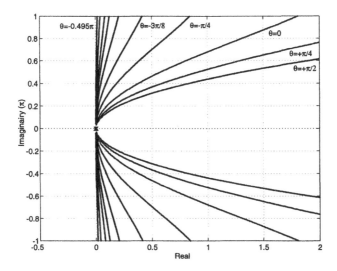

Fig. 3-17 *Root locus of a second-order $\Sigma\Delta$ modulator for several quantizer phases θ at $f_s/2$*

The characteristic equation of the second-order system Fig. 3-15 now becomes

$$1 + 2\kappa e^{s\theta}\left(\frac{1 - e^{-sT_s}}{s}\right) \cdot \left(\frac{s + 0.5}{s^2}\right) = 0 \qquad (3-29)$$

The root locus plot of Fig. 3-18 shows that the poles move into the right-half plane with increasing quantizer gain κ. The poles move back into the left-half plane if the gain decreases. This may result in a stable periodic sequence, or limit cycle, at the output of the modulator. A limit cycle occurs if the

- *closed-loop gain is 1*
- *phase shift in the loop is 2π*

The gain and phase shift of the loopfilter and the zero-order hold function are determined for all frequencies. Therefore, a limit cycle exists if there is a solution for the gain and the phase of the quantizer such that the criteria above are met. It was observed from Fig.3-16 that the quantizer phase range is large at $f_s/4$ ($s=j\omega_s/4$). At $f_s/4$, the filter gain and phase are 1.336 and 0.6π, while the gain and phase shift of the zero-order hold function are 0.9 and $\pi/4$ respectively. Taking into account the sign inversion of the feedback path (phase π), the quantizer phase shift must be

$$\theta = 2\pi - \pi - \frac{\pi}{4} - 0.6\pi = 0.15\pi \qquad (3\text{-}30)$$

0.15π at $f_s/4$ (equal to 0.3π at $f_s/2$) and the quantizer gain

$$\kappa = \frac{1}{0.9 \cdot 1.336} = 0.83 \qquad (3\text{-}31)$$

for a limit cycle at $f_s/4$.

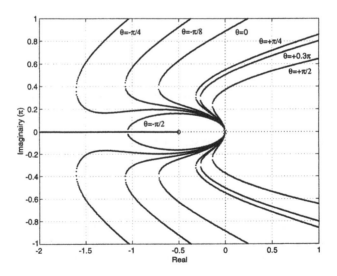

Fig. 3-18 *Root locus of a second-order $\Sigma\Delta$ modulator with a zero in the loopfilter transfer function as a function of the quantizer gain κ, for several quantizer phases θ at $f_s/2$*

Fig. 3-18 shows the root locus plot of the second-order modulator for several quantizer phases. At a quantizer phase of 0.15π at $f_s/4$ (0.3π at $f_s/2$) the root locus enters the right-half plane at $f_s/4$. The quantizer gain at this intersection point is equal to the result of Eq. (3-31). Fig. 3-19 shows the simulated output spectrum of the second-order loopfilter with the transfer function of Eq. (3-28), in the case of a very small input signal. In this simulation the limit cycle at $f_s/4$ is dominant.

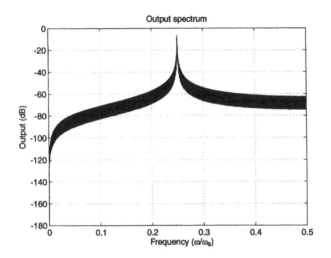

Fig. 3-19   *Output spectrum of a second-order ΣΔ modulator for a very small input signal*

3.3.2 Large signal stability

In the previous section, stability of a second-order ΣΔ modulator for small input signals was discussed. If the amplitude of the input signal increases, the quantizer gain decreases and the poles of a higher-order modulator may eventually be pushed into non-stable locations. The reason for this instability is that the input signal becomes too large to be compensated for by the feedback signal. Typically, the maximum specified input signal ranges from 50% to 80% of the rms-value of the DAC feedback signal [3.2]. Large signal instability also becomes visible from a root locus plot of a fifth-order modulator in Fig. 3-21 (example quantizer phase is 0.3π at $f_s/2$). The loopfilter of this fifth-order modulator has 5 poles and four zeros. If the quantizer gain drops below κ_s two poles shift into the right-half plane and the system becomes unstable. The gain κ_s determines the maximum allowable input signal level for a stable modulator.

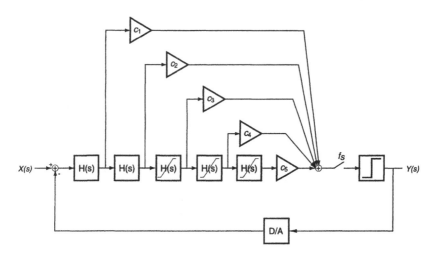

Fig. 3-20 *Fifth-order $\Sigma\Delta$ modulator with limiters in the higher-order integrator stages for overload stability*

An effective method to preserve the stability of a $\Sigma\Delta$ modulator at overload is to limit the maximum output swings of the loopfilter integrator stages. By limiting the integrator swings, the quantizer gain is prevented from dropping below the stable value κ_s. Fig. 3-20 shows a fifth-order modulator with limiters in the third, fourth and fifth filter stages. The feedforward paths with gains c_1-c_5 are needed to create the zeros in the loopfilter for stability of the loop (section 3.4.1). The filter coefficients should be scaled in such a way that, at an increasing overload input level, the integrator outputs are limited one by one, starting with the last integrator stage. As a result, the modulator order is degraded from fifth to fourth, to third and finally a second-order modulator remains which is always stable for large input signals (Fig. 3-18). If the input level drops from the overload state to a lower stable value again, stable fifth-order behavior is retrieved. Another technique to ensure large signal stability is to monitor the number of consecutive equal bits in the digital domain. In general, the maximum number of equal bits in a row is in the order of 10 bits for a stable modulator which is not in overload. If the number of bits becomes too large the integrator outputs are reset, by switching the integrator outputs to a zero reference. The resetting of the integrator states continues as long as the modulator remains in overload.

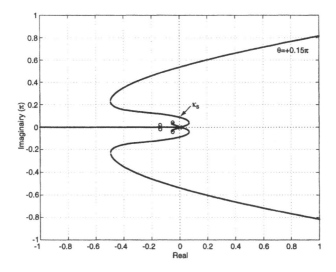

Fig. 3-21 *Root locus of a fifth-order ΣΔ modulator with poles and four zero in the loopfilter transfer function (example quantizer phase is 0.15π at $f_s/4$). If the quantizer gain drops below κ_s, the system becomes unstable.*

3.4 High-frequency stability compensation

In this section, the design of higher-order filters is presented. Feedforward and feedback compensation techniques are described for high frequency stability of the loop. Both compensation techniques can provide the same noise transfer function but have different signal transfer characteristics. Butterworth and inverse Chebyshev noise transfer functions are shown. Finally, a comparison is made between the two compensation techniques.

3.4.1 Feedforward compensation

Increasing the order of the loopfilter of a ΣΔ modulator provides more aggressive quantization noise-shaping and efficiently improves signal-to-quantization noise ratio (section 3.2.1).

The general form of a nth-order integrator filter is given by

$$H(s) = \left(\frac{\omega_u}{s}\right)^n \tag{3-32}$$

Where ω_u is the unity-gain frequency of all integrators. However, a sigma-delta modulator with the filter transfer of Eq. (3-32) is not stable for n larger than 1. An uncompensated second-order loopfilter has -180 degrees phase shift at high frequencies. This amount of phase shift is too large for stable idling and a zero should be introduced in the loopfilter transfer function to reduce the high frequency phase shift. This can be done by adding a feedforward path in the loopfilter.

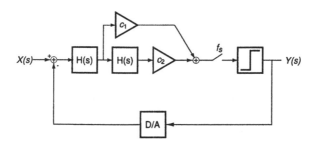

Fig. 3-22 *Second-order* $\Sigma\Delta$ *modulator with feedforward compensation*

The topology of a second-order modulator with feedforward compensation is shown in Fig. 3-22. The choice for the ratio between c_2 and c_1 is determined by considerations concerning stability, maximum input level, signal-to-noise ratio and parameter spread due to non-ideal processing. By increasing the coefficient ratio c_1/c_2, the zero in the transfer function is shifted to a lower frequency. An optimum can be found which provides both high frequency stability and second-order noise shaping at low frequencies. The choice for the coefficient value c_1 cannot be too critical to prevent process spread from pushing the modulator to instability. Using the quantizer model as described in Eq. (3-24), the characteristic equation of the system of Fig. 3-22 is

$$1 + \kappa e^{s\theta}\left(\frac{1 - e^{-sT_s}}{s}\right)\frac{\left(c_1 \cdot \frac{1}{\omega_u}s + c_2\right)}{\left(\frac{1}{\omega_u}s\right)^2} = 0 \tag{3-33}$$

The zero of Eq. (3-33) equals

$$z = -\frac{c_2}{c_1} \cdot \omega_u \qquad (3\text{-}34)$$

The root locus plot was shown in Fig. 3-18. Indeed, Eq. (3-34) shows that by increasing the value of c_1 relative to c_2, the loopfilter zero moves to lower frequencies which improves system stability at the cost of less effective noise-shaping (first-order behavior). Decreasing the value of c_2 results in a more aggressive way of shaping quantization noise by means of a more critical stable system (second-order behavior). Common values for c_1 and c_2 are 2 and 1 respectively [3.13], resulting in a zero of

$$z = -0.5\omega_u \qquad (3\text{-}35)$$

Next, the design of a fifth-order loopfilter will be explored. The same analysis as before can be applied. However, the fifth-order filter topology offers more design freedom to increase the SNR. First, the design of a Butterworth filter is elaborated. Then, it is shown that by adding local feedback paths, to create an inverse Chebyshev NTF, even higher resolution is achieved.

Figure 3-23 shows the fifth-order $\Sigma\Delta$ modulator model with feedforward compensation. The loopfilter consists of five single-pole integrator stages with transfer functions conform to Eq. (3-32). The integrator stages have unity-gain frequencies ω_1-ω_5 and feedforward coefficients c_1-c_5. The transfer function of a fifth-order modulator with feedforward compensation is

$$STF = \frac{\kappa e^{s\theta} H(s)}{\dfrac{s^5}{\omega_1\omega_2\omega_3\omega_4\omega_5} + \kappa e^{s\theta}\left(\dfrac{1-e^{-sT_s}}{s}\right)H(s)} \qquad (3\text{-}36)$$

where $H(s)$ is

$$H(s) = c_5 + \frac{s}{\omega_5} \cdot \left(c_4 + \frac{s}{\omega_4} \cdot \left(c_3 + \frac{s}{\omega_3} \cdot \left(c_2 + c_1 \cdot \frac{s}{\omega_2}\right)\right)\right) \qquad (3\text{-}37)$$

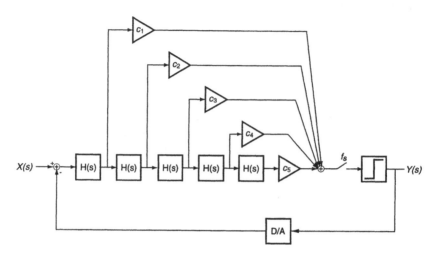

Fig. 3-23 *Fifth-order* $\Sigma\Delta$ *modulator with feedforward compensation*

For high frequencies the STF roll-off is first-order. In general this implies that for high frequency input signals, the feedforward compensated loopfilter acts as a first-order anti-aliasing filter. The system characteristic equation is

$$1 + \kappa e^{s\theta}\left(\frac{1 - e^{-sT_s}}{s}\right)\frac{c_5 + \dfrac{s}{\omega_5}\cdot\left(c_4 + \dfrac{s}{\omega_4}\cdot\left(c_3 + \dfrac{s}{\omega_3}\cdot\left(c_2 + c_1\cdot\dfrac{s}{\omega_2}\right)\right)\right)}{s^5/(\omega_1\omega_2\omega_3\omega_4\omega_5)} = 0 \qquad (3\text{-}38)$$

To get a maximally flat transfer function within the signal band, the zeros are placed in Butterworth position. The Butterworth positions for a fourth-order system are

$$s_k = \omega_z\cdot e^{j\pi\left(\frac{-3 + 2k}{8}\right)}, k = 0, 1, 2, 3 \qquad (3\text{-}39)$$

The general form of a fourth-order Butterworth filter is

$$((s + \omega_z\cdot\cos\varphi_1) + \omega_z^2\cdot\sin^2\varphi_1)\cdot((s + \omega_z\cdot\cos\varphi_2) + \omega_z^2\cdot\sin^2\varphi_2) \qquad (3\text{-}40)$$

where φ_1 and φ_2 are the angles of the zeros in the complex plane relative to the imaginary axis, which are calculated by Eq. (3-39). In order to place the zeros of the characteristic equation in Butterworth positions Eq. (3-37) and Eq. (3-40) should be equal. This results in five relations to calculate the desired coefficients

$$c_1 = a_1 \cdot \frac{\omega_2 \omega_3 \omega_4 \omega_5}{\omega_z^4} \tag{3-41}$$

$$c_2 = a_2 \cdot \frac{\omega_3 \omega_4 \omega_5}{\omega_z^3}$$

$$c_3 = a_3 \cdot \frac{\omega_4 \omega_5}{\omega_z^2}$$

$$c_4 = a_4 \cdot \frac{\omega_5}{\omega_z}$$

$$c_5 = a_5$$

or, in general,

$$c_m = a_m \cdot \frac{\displaystyle\prod_{i = m+1}^{n} \omega_i}{(\omega_z)^{n-m}}, \quad m = 1, 2, \dots n \tag{3-42}$$

where a_m is the Butterworth coefficient and n the loopfilter order. The Butterworth coefficients a_1-a_5 of a fifth-order filter are

$$a_1 = 1 \tag{3-43}$$

$$a_2 = 2 \cdot \left(\cos\left(\frac{1}{8}\pi\right) + \cos\left(\frac{3}{8}\pi\right) \right)$$

$$a_3 = 4 \cdot \left(\cos\left(\frac{1}{8}\pi\right) \cdot \cos\left(\frac{3}{8}\pi\right) + \frac{1}{2} \right)$$

$$a_4 = 2 \cdot \left(\cos\left(\frac{1}{8}\pi\right) + \cos\left(\frac{3}{8}\pi\right) \right)$$

$$a_5 = 1$$

Figure 3-24 shows an example of a Bode plot of the fifth-order filter. At low frequencies the filter characteristic rolls off fifth-order and the phase shift closely reaches -450 degrees (at 10^{-2} radians). At high frequencies the filter becomes first-order and the phase turns back to about -97 degrees at half the sampling frequency. In general, any order of filter can be designed according to the procedure as discussed in this section. However, further increasing the filter order pushes the zero locations to lower frequencies for a stable system. Hence, increasing the filter order decreases the effective bandwidth where quantization noise is shaped to a low enough level. Because of the fifth-order shaping, low frequency quantization noise is very small and rises fifth-order with frequency. Therefore, total in-band quantization noise is dominantly determined by the quantization noise at the end of the signal band. This severely limits the maximum obtainable bandwidth. An effective way to shape the end-of-the-band quantization noise is to apply local feedback paths in the loopfilter which create resonator stages. The resonator stages can provide some extra notch gain at the end of the signal band which suppresses the large amount of quantization noise that is present here. The design of a loopfilter with local feedback paths will be discussed next.

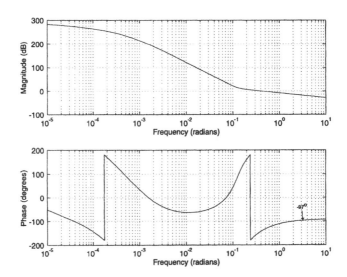

Fig. 3-24 *Bode plot of fifth-order loopfilter with feedforward compensation*

3.4.2 Feedforward compensation and local feedback

Applying feedback around two integrator stages shifts the DC gain of the integrators to a finite resonance frequency. This way, the gain can be spread more equally in the signal band which, for example, is important for suppression of quantization noise at the end of the signal band. The general transfer function of a resonator is

$$H(s) = \frac{\omega_u s}{s^2 + \omega_u^2} \tag{3-44}$$

With local feedback paths it is possible to implement an inverse Chebyshev NTF rather than a Butterworth NTF. Fig. 3-25 shows the general model of a fifth-order modulator with feedforward and local feedback paths.

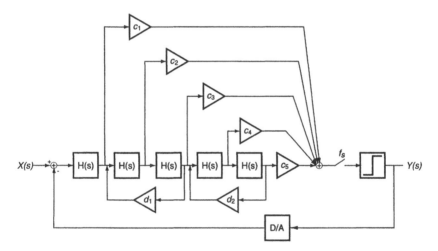

Fig. 3-25 *Fifth-order baseband $\Sigma\Delta$ modulator with feedforward compensation and local feedback*

The feedforward coefficients and the integrator unity gain frequencies remain unchanged while the local feedback coefficients d_1 and d_2 determine the notch frequencies of the resonators according to

$$\omega_{n1} = \sqrt{\omega_2 \omega_3 d_1}, \ \omega_{n2} = \sqrt{\omega_4 \omega_5 d_2} \tag{3-45}$$

Fig. 3-26 shows the NTF plot of the modulators of Fig. 3-23 and Fig. 3-25 respectively. The resonator gain notches provide a relatively flat quantization noise spectrum in the signal band which greatly improves the signal-to-quantization-noise ratio (SQNR) compared to the fifth-order loopfilter without local feedback. The high frequency filter behavior is hardly affected and the system remains stable. Comparing both NTF plots reveals that the DC gain of the filter with local feedback is smaller than that of the filter without local feedback. However, the quantization noise power is not dominantly determined by DC gain, but by the gain at the edge of the signal band. Therefore, the inverse Chebyshev filter provides a more efficient noise-shaping function for optimizing SQNR. Moreover, DC tones (Fig. 3-5) and harmonic components of the rectangular feedback signal (Fig. 3-10) are attenuated as a result of the extra gain at the end of the signal band.

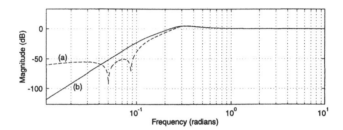

Fig. 3-26 *NTF of fifth-order ΣΔ modulator with feedforward compensation with (a) local feedback (dashed line), and (b) without local feedback (solid line).*

3.4.3 Feedback compensation

In the previous sections, feedforward compensation was used for stabilization of higher-order modulators. An alternative method is to use feedback paths as shown in Fig. 3-27 for higher-order filters. A fraction of the output is fed to the input of each integrator stage in the loopfilter. Using the linear quantizer model of Fig. 3-15, the signal transfer function of a second-order modulator with feedback compensation is

$$STF = \frac{\dfrac{\omega_1 \omega_2}{s^2} \kappa e^{s\theta}}{1 + \kappa e^{s\theta}\left(\dfrac{1 - e^{-sT_s}}{s}\right)\left(c_2 \dfrac{\omega_2}{s} + c_1 \dfrac{\omega_1 \omega_2}{s^2}\right)} \qquad (3\text{-}46)$$

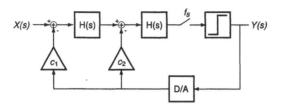

Fig. 3-27 *Second-order $\Sigma\Delta$ modulator with feedback compensation*

The characteristic equation is equal to the denominator of Eq. (3-47)

$$1 + \kappa e^{s\theta}\left(\frac{1 - e^{-sT_s}}{s}\right)\left(\frac{c_2\omega_2 s + c_1\omega_1\omega_2}{s^2}\right) = 0 \qquad (3\text{-}47)$$

This relation is similar to the characteristic equation of the feedforward compensated system of Fig. 3-22.

3.4.4 Feedback compensation and local feedback

Fig. 3-28 shows a fifth-order $\Sigma\Delta$ modulator with feedback compensation. The STF of this modulator is

$$STF = \frac{\kappa e^{s\theta}}{\dfrac{s^5}{\omega_1\omega_2\omega_3\omega_4\omega_5} + \kappa e^{s\theta}\left(\dfrac{1 - e^{-sT_s}}{s}\right)(b_4 s^4 + b_3 s^3 + b_2 s^2 + b_1 s + c_1)} \qquad (3\text{-}48)$$

where b_1-b_4 are coefficients

$$b_n = c_{n+1} \prod_{m=1}^{n} \frac{1}{\omega_m}, n = 1, 2, 3, 4 \qquad (3\text{-}49)$$

As has been the case for the second-order filter, the STF of the fifth-order filter with feedback compensation does not have zeros (assuming θ is 0). Thus, the STF of a nth-order feedback compensated $\Sigma\Delta$ modulator is a nth-order lowpass filter. The characteristic equation equals

$$1 + \kappa e^{s\theta}\left(\frac{1 - e^{-sT_s}}{s}\right)\left(\frac{b_4 s^4 + b_3 s^3 + b_2 s^2 + b_1 s + c_1}{s^5/(\omega_1 \omega_2 \omega_3 \omega_4 \omega_5)}\right) = 0 \qquad (3\text{-}50)$$

For a maximal flat filter transfer, the coefficients c_1-c_5 are placed in Butterworth positions, according to Eq. (3-40).

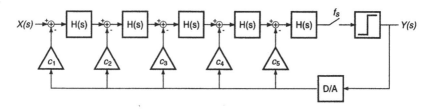

Fig. 3-28 *Fifth-order $\Sigma\Delta$ modulator with feedback compensation*

With local feedback paths, the Butterworth NTF can be changed into an inverse Chebyshev NTF, to optimize the noise-shaping function. Fig. 3-29 shows the feedback compensated topology with local feedback. Again, the filter coefficients are determined following the same method as in section 3.4.2.

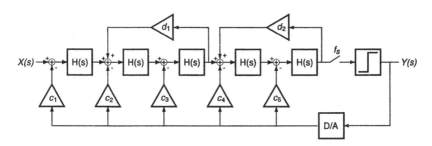

Fig. 3-29 *Fifth-order $\Sigma\Delta$ modulator with feedback compensation and local feedback*

3.4.5 Feedforward versus feedback compensation

It has been shown that compensation of a higher-order loopfilter with feedforward or feedback paths, yields similar relations for the characteristic equation. Thus, for stability of the loop, both compensation techniques can be used successfully. Nevertheless, some essential differences exist between the two topologies.

The STFs of the feedforward and feedback compensated loopfilters are given by Eq. (3-36) and Eq. (3-48) respectively. Generally, the STF of an nth-order feedforward compensated filter has n poles and n-1 zeros, while the feedback compensated STF has n poles only. Therefore, the feedforward STF is a first-order lowpass anti-aliasing filter, while the feedback STF is an nth-order lowpass anti-aliasing filter. This is shown graphically in Fig.3-30 for the fifth-order modulators as described in section 3.4.1 and section 3.4.3. Clearly, the feedback STF provides much stronger filtering for high frequency signals than the feedforward STF. A second difference between the STFs can be observed from Fig.3-30 as well. It is shown that, due to the zeros of the feedforward filters, the STF is not flat at low frequencies for the choice of filter coefficients as derived earlier. Because of the non-ideal compensation of STF poles and zeros, the STF shows peaking at a certain frequency. This implies that at the peaking frequency the maximum stable input level is reduced by the gain of the peaking. Because the STF has no zeros, the feedback filter does not suffer from significant peaking.

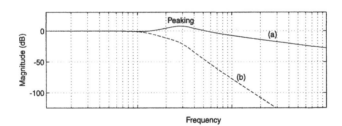

Fig. 3-30 *STF of fifth-order (a) feedforward compensated loopfilter (solid) and (b) feedback compensated loopfilter (dashed)*

A major drawback of the feedback filter architecture is that the outputs of the integrators contain, besides filtered quantization noise, a substantial part of the input signal [3.2]. The unity gain frequencies of the lower order integrators (especially the first integrator) in a feedback filter have to be much lower than in the case of a feedforward filter, in order to keep the output signal swing within the available supply range. Thus, noise and distortion of the higher-order integrators can not be neglected as they are not heavily suppressed by a high gain of the previous integrator stages. This can be a serious problem, especially if the oversampling ratio is low. As a result, power consumption of the feedback architecture tends to be higher than that of a feedforward architecture. This is the main reason why feedforward filters have been used for the designs in this book. It should be noted that by combining feedforward and feedback compensation techniques, more freedom is obtained for an optimal filter design.

3.5 Quadrature sigma-delta modulation

In the previous section, the design of higher-order lowpass filters was discussed for application in a $\Sigma\Delta$ modulator with real input signals. The poles and zeros of the filter are real and complex-conjugate pairs, due to the real-valued filter coefficients. If two $\Sigma\Delta$ modulators with real filters are used in a quadrature configuration (Fig. 2-7), with complex input $I + jQ$, the magnitude response is symmetrical around DC. By using complex-valued coefficients, the filter transfer is not constrained to having complex-conjugate pairs of poles or zeros [3.20] and it can be asymmetrical around DC. This will be shown for a complex integrator in section 3.5.1. Then, the design of a complex fifth-order filter is shown as an example. Again, both feedforward and feedback frequency compensation techniques can be used in a complex filter for stability.

3.5.1 Complex integrator

A complex integrator consists of two real integrator stages, with quadrature input signals, which are cross-coupled. The model of a continuous-time complex integrator is shown in Fig. 3-31. The transfer $H_c(s)$ of the complex integrator equals

$$H_c(s) = \frac{A}{\dfrac{A}{\omega_u}s + 1 + jeA} \tag{3-51}$$

The transfer of Eq. (3-51) is almost identical to the transfer of a real integrator, with the difference that the frequency independent part in the denominator is a complex number. The coefficient e causes a shift in the frequency domain of the real transfer $H(s)$ into the position of the complex transfer $H_c(s)$. This is shown graphically in Fig. 3-32 for the NTF of a first-order $\Sigma\Delta$ modulator with the filter transfer $H_c(s)$ of Eq. (3-51), for e is 0, e is +0.1 and e is -0.2. If e is zero, the complex transfer $H_c(s)$ is equal to the real transfer $H(s)$ and the NTF is symmetrical around DC with the gain notch at DC (Fig. 3-32a). If e is not equal to zero, the NTF shifts from DC to $e\omega_u$. The unity gain frequency ω_u' of $H_c(s)$ is equal to

$$\omega'_u = \omega_u \cdot \sqrt{1 + e^2} \qquad (3\text{-}52)$$

For small values of e ($e < 0.1$) the unity-gain frequency only slightly changes relative to the unity-gain frequency of a real integrator. This implies that stability of a $\Sigma\Delta$ modulator is hardly affected in case the loopfilter consists of complex integrators. If the value of e becomes closer to one, the stability coefficients must be recalculated with the unity gain frequency of Eq. (3-52). If the feedback coefficient e has a negative value, the gain notch of the complex transfer function $H_c(s)$ shifts into the negative frequency plane. Fig. 3-32b and Fig. 3-32c show the NTFs in the case e has a positive and a negative value respectively. This way, it is possible to shift the gain notches of all complex integrators in a nth-order complex filter into the positive (or negative) frequency plane for strong quantization noise suppression. This is shown in the next section.

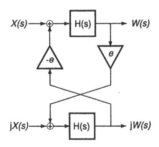

Fig. 3-31 *Complex integrator filter*

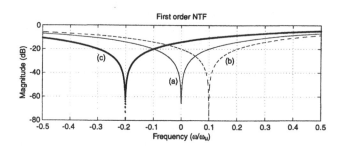

Fig. 3-32 *NTF of first-order complex filter with complex local feedback coefficient e equal to (a) 0, (b) +0.1, (c) -0.2*

3.5.2 Complex filter design

Fig. 3-33 shows the model of a fifth-order quadrature baseband $\Sigma\Delta$ modulator with feedback compensation. The gain notches of the 5 complex integrators can be shifted into the positive frequency plane and distributed within the signal band. Such a fifth-order NTF of a quadrature modulator is shown in Fig. 3-34. The first notch of the NTF is at DC, which means that one of the complex cross-coupling coefficients is equal to zero. The notches are positioned such, that the NTF is flat within the signal band. With 5 notches, a deep suppression of quantization errors within the signal band is obtained. Outside the signal band, the NTF rises fifth-order.

An ideal fifth-order complex baseband $\Sigma\Delta$ modulator with the NTF characteristic of Fig. 3-34 has been simulated and the bitstream output spectrum is shown in Fig. 3-36. As expected, quantization noise within the signal band is very low and the theoretical obtainable SQNR is high. In practice, the SQNR will be impaired by non-ideal effects. The most important error sources are process related, which are parameter mismatch and spread. As we are dealing with a complex system, the problem of image interference arises, as explained in section 2.4.3. Because of the asymmetrical NTF, the quantization noise in the image band is not suppressed, as can be seen from Fig. 3-36. Therefore, quantization noise power in the image band is very large. In case of a mismatch between the quadrature feedback signals in a complex modulator, part of the quantization noise power in the image band 'leaks' into the signal band and vice versa [3.20]- [3.22]. This effect has been simulated and Fig. 3-37 shows the output spectrum of the complex $\Sigma\Delta$ modulator in the case of 0.5% mismatch between the quadrature feedback signals. Because of the high quantization noise power in the image band, the SQNR can be affected considerably, as a result of image interference.

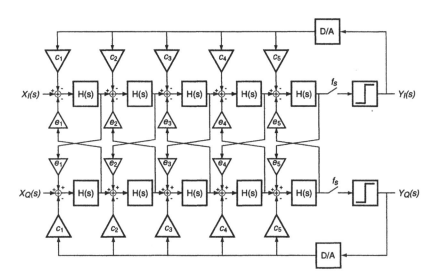

Fig. 3-33 *Fifth-order quadrature baseband $\Sigma\Delta$ modulator with feedback compensation*

To reduce the effect of quantization noise leakage from the negative band in the positive band, the matching between the quadrature signals should be improved. A powerful technique to improve matching is dynamic element matching. This is discussed in detail in chapter 5. A second method is to reduce the quantization error power in the image band. This can be done by shifting one integrator notch in the image band at the cost of less effective quantization error suppression in the signal band [3.20]. The NTF of the fifth-order complex $\Sigma\Delta$ modulator is shown in Fig. 3-35. Because quantization noise power is largest at the end of the image band, the integrator notch is placed close to the end of the band. Because of the loss of an integrator notch, the quantization noise suppression in the signal band is lower than in Fig. 3-34.

Process spread is another non-ideality which can seriously affect SQNR performance. This problem especially applies to the complex $\Sigma\Delta$ modulator topology because of the very high in-band quantization noise suppression. Any frequency shift of the integrator notches, in particular the notch at the end of the signal band, deteriorates SQNR. The filter coefficients and unity-gain frequencies rely on absolute component values rather than ratios. Due to non-ideal processing, the absolute value can spread with 10-20%. Therefore, coefficient calibration is highly recommended for this kind of filter

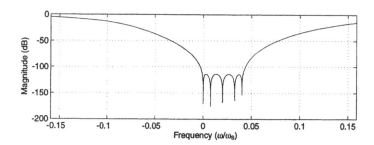

Fig. 3-34 *NTF of fifth-order complex ΣΔ modulator with 5 complex integrators. All integrator gain notches are positioned in the positive frequency plane*

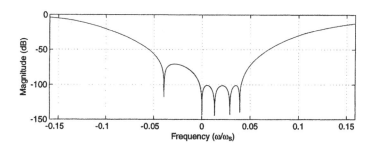

Fig. 3-35 *NTF of fifth-order complex ΣΔ modulator with 5 complex integrators. One integrator gain notch is positioned in the negative frequency plane*

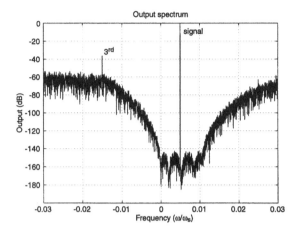

Fig. 3-36 *Simulated output spectrum of an ideal fifth-order quadrature baseband ΣΔ modulator*

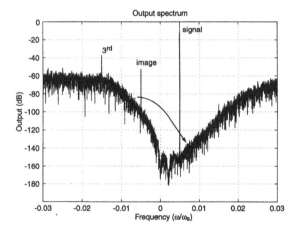

Fig. 3-37 *Simulated output spectrum of a fifth-order quadrature baseband ΣΔ modulator with 0.5% gain mismatch between the quadrature feedback signals*

References

[3.1] Cutler, C.C., "Transmission system employing quantization," U.S. Patent No. 2,927,962, March 8, 1960 (filed 1954).

[3.2] Norsworthy, S.R., R. Schreier, G.C. Temes, *Delta-sigma data converters - Theory, design, and simulation*, IEEE Press, New York, 1997.

[3.3] Inose, H., Y. Yasuda, and J. Murakami, "A telemetering system by code modulation-Δ-Σ modulation," *IRE Trans. Space Electron. Telemetry*, vol. SET-8, pp. 204-209, Sept. 1962.

[3.4] Jager, F. de, "Delta modulation - a method of PCM transmission using the one unit code," *Philips Res. Rep.*, vol. 7, pp. 442-466, 1952.

[3.5] Bennett, W.R., "Spectra of quantized signals," *Bell Syst. Tech. J.*, vol. 27, pp. 446-472, Jul. 1948.

[3.6] Zwan, E.J. van der, E.C. Dijkmans, "A 0.2 mW CMOS $\Sigma\Delta$ modulator for speech coding," *IEEE J. Solid-State Circuits*, vol. 31, pp. 1873-1880, Dec. 1996.

[3.7] Schuchman, L., "Dither signals and their effect on quantization noise," *IEEE Trans. Commun. Tech.*, vol. COM-12, pp. 162-165, Dec. 1964.

[3.8] LaMay, J.L., and H. T. Bogard, "How to obtain maximum practical performance from state-of-the-art delta-sigma analog-to-digital converters," *IEEE Trans. Instr. Meas.*, vol. 41, pp. 861-867, Dec. 1992.

[3.9] Breems, L.J., E.J. van der Zwan, E.C. Dijkmans and J.H. Huijsing, "A 1.8 mW CMOS $\Sigma\Delta$ Modulator with Integrated Mixer for A/D Conversion of IF Signals," *ISSCC Dig. Tech. Papers*, pp. 64-65, Feb. 1999.

[3.10] Breems, L.J., E.J. van der Zwan, J.H. Huijsing, "Design for Optimum Performance-to-Power Ratio of a Continuous-time $\Sigma\Delta$ Modulator," *Proc. of ESSCIRC*, pp. 318-321, Sep. 1999.

[3.11] Adams, R.W., "Design and implementation of an audio 18-bit analog-to-digital converter using oversampling techniques," *J. Audio Eng. Soc.*, vol. 34, pp. 153-166, Mar. 1986.

[3.12] Westra, J.R., *High-performance oscillators and oscillator systems*. Ph.D. thesis, Delft University of Technology, 1998.

[3.13] Plassche, R. van der, *Integrated analog-to-digital and digital-to-analog converters*, Kluwer Academic Publishers, Dordrecht, 1994.

[3.14] Kirk, C.-H, Chao, Shujaat, N., Wai L. Lee, and Charles G. Sodini, "A higher order topology for interpolative modulators for oversampling A/D converters," *IEEE Trans. on Circuits and Systems*, vol. 37, pp. 309-318, Mar. 1990.

[3.15] Bhagawati, P.A., K. Shenoi, "Design methodology for $\Sigma\Delta$M," *IEEE Trans. on Communications*, vol. 31, pp. 360-370, Mar. 1983.

[3.16] Atherton, D.P., *Stability of non-linear systems*, Research Studies press; Wiley, ISBN 0-387-94582-2, 1981.

[3.17] Ardalan, S.H., and J.J. Paulos, "Analysis of nonlinear behavior in delta-sigma modulators," *IEEE Trans. Circuits Sys.*, vol. 34, pp. 593-603, June 1987.

[3.18] Hoffelt, M.H., "On the stability of a 1-bit quantized feedback system," *Proc. of ICASSP*, pp. 844-848, 1979.

[3.19] Engelen, J. van, *Stability analysis and design of bandpass sigma delta modulators*, Ph.D. thesis, Technische Universiteit Eindhoven, 1999.

[3.20] Jantzi, S.A., K.W. Martin, A.S. Sedra, "Quadrature bandpass $\Delta\Sigma$ modulation for digital radio," *IEEE J. Solid-State Circuits*, vol. 32, pp. 1935-1950, Dec. 1997.

[3.21] Jantzi, S.A., K. Martin, A.S. Sedra, "A quadrature bandpass delta-sigma modulator for digital radio," *ISSCC Dig. Tech. Papers*, pp. 216-217, Feb. 1997.

[3.22] Jantzi, S.A., K.W. Martin and A.S. Sedra, "Mismatch effects in complex bandpass $\Delta\Sigma$ modulators," *Proc. of ISCAS*, vol. 1, pp. 227-230, May 1996.

Realization of an IF-to-baseband sigma-delta modulator

4

4.1 Introduction

In this chapter the theory, design, and experimental results of a continuous-time $\Sigma\Delta$ modulator with an integrated mixer are presented. This IF-to-baseband $\Sigma\Delta$ modulator has the functions of a mixer, anti-aliasing filter, and A/D converter. Key features of this topology are its low-power consumption (1.8 mW) and high linearity (-84 dB IM3). In section 4.2 a general overview is given of two different $\Sigma\Delta$ modulator architectures which incorporate frequency translation inside and outside the $\Sigma\Delta$ loop. It is shown that non-idealities of the mixer can result in modulation of high-frequency quantization noise into the frequency band of interest. In the final part of the section, the topologies are compared. In section 4.3, the integrated design of a passive mixer and an RC filter is shown that is used for the input stage of the IF $\Sigma\Delta$ modulator. Important aspects concerning linearity, mixer driving and self-mixing are discussed. Mathematical derivations give 'rule-of-thumb' relations between design parameters, such as biasing current and transistor dimensions, and performance figures, like noise and distortion. In section 4.4 the complete $\Sigma\Delta$ modulator topology is presented and the design of the individual blocks are shown. The experimental results of two realized prototype chips are given in section 4.5. In section 4.6 conclusions are drawn.

4.2 Frequency translation in sigma-delta modulators

In this section, two topologies for IF-to-baseband conversion in a ΣΔ modulator are shown which incorporate a mixer inside (section 4.2.1) and outside (section 4.2.2) the loop of a baseband ΣΔ modulator. An important requirement is that an independent choice can be made for the LO frequency of the mixer and the sampling frequency of the ΣΔ modulator. If not related, the sampling rate can be optimized for dynamic range performance and power consumption of the ΣΔ modulator, without affecting the LO frequency, which is often set by the application. In section 4.2.3, both topologies are compared and conclusions are drawn.

4.2.1 Mixer inside the sigma-delta loop

In this section ΣΔ modulator topologies are investigated with the mixer inside the sigma-delta loop. Fig. 4-1 shows a general model of a single-loop ΣΔ modulator with a mixer inside the loop. The IF input signal X is mixed with the LO frequency (ω_{LO}) and translated to baseband by the mixer in the forward path (mixer1 in Fig. 4-1). The baseband signal is further processed by a lowpass loopfilter and digitized. The feedback signal needs to be upconverted by a second mixer (mixer2 in Fig. 4-1) to compensate for the downconversion mixer in the forward path. The input signal X is adjusted by the gain of the mixer in the forward path only, while the quantizer error passes both mixers in the loop and its gain is not affected. Therefore, the maximum allowable input level (non-overload) is adjusted by the gain of the mixer. The stability of the ΣΔ modulator is not affected, because the two mixers in the loop cancel each other.

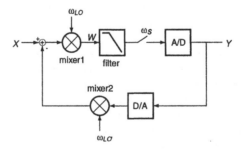

Fig. 4-1   *Model of a continuous-time ΣΔ modulator with mixer inside the loop*

The main advantage of placing the mixer inside the ΣΔ loop is that non-idealities of the mixer in the forward path (mixer1), like non-linearity, are suppressed by the loopgain. This is the case for ΣΔ modulators with a multi-bit DAC (section 3.2.3) or if the feedback path incorporates an analog filter [4.1]. If a ΣΔ modulator has a single-bit DAC without a filter in the feedback path, the mixer non-idealities are not suppressed by the loop. Note that the non-linearity of the mixer in the feedback path (mixer2) is also not suppressed by the loop. If the LO frequency is a rational fraction of the sampling frequency, the frequency translation of the feedback signal (mixer2) can be easily performed in the digital domain with perfect linearity.

Due to the continuous-time nature of the signals, the continuous-time system of Fig.4-1 is sensitive to parasitic delay between the mixers. If mixing of the feedback signal is done in the digital domain, parasitic delay in the DAC adds to the total delay. A delay between the two mixers can be modeled as a delay between the LO signals of both mixers. This is shown in the time domain in Fig.4-2a, which shows delayed clock signals *LO* (mixer1) and *LO'* (mixer2), the output signal *Y*, and the mixer1 output *W*. In this example, the LO frequency is equal to half the sampling frequency. During a small time interval the output signal *Y* passes the mixers in the feedback and feedforward path, while being modulated only once. In the frequency domain, this implies that a fraction of the noise band around the LO carrier in the output bitstream spectrum is folding into the signal band, which can deteriorate system performance.

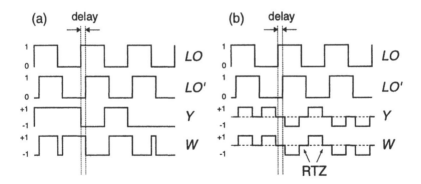

Fig. 4-2   *Effect of phase delay between LO signals, on (a) full period continuous-time DAC pulse; (b) continuous-time DAC pulse with RTZ*

This effect has been simulated in the case of a delay of 0.1% of the LO period between the LO signals of both mixers (Fig.4-3a). Again, the LO frequency is chosen equal to half the sampling frequency (ω_s). A -20 dB sinusoidal signal, modulated at the LO frequency with a small offset frequency of $0.001\omega_s$, has been applied to the input of the system of Fig.4-1. Fig.4-3b shows the same simulation, without any delay. Clearly, the noise floor is considerably higher in case of a small delay, compared to the ideal simulation. This effect is considerably reduced if the LO frequency is close to the sampling frequency. Around the sampling frequency, quantization noise is greatly suppressed by the loopfilter, and noise leakage into baseband will be much less severe.

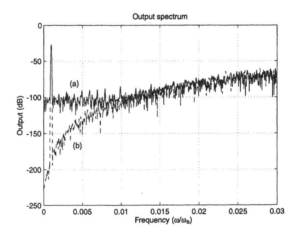

Fig. 4-3   *Simulated output spectrum of continuous-time $\Sigma\Delta$ modulator with mixers inside the loop, (a) with 0.1% delay between mixer LO frequencies ($\omega_{LO} = \omega_s/2$); (b) without any delay*

A method to overcome the delay problem is the use of DAC pulses with return-to-zero. Fig.4-2b again shows the LO waveforms of the mixers and the DAC signal, now with RTZ. Due to the RTZ intervals, the DAC is insensitive to a delay between the LO signals as long as the delay falls within the RTZ interval. This way the LO frequency can be a rational fraction of the sampling frequency, without performance degradation. Note that a discrete-time (switched-capacitor) $\Sigma\Delta$ modulator does not have the delay problem as the input and feedback signals are sampled and almost perfectly synchronized.

So far, single-loop $\Sigma\Delta$ modulators have been shown with a mixer inside the loop. Fig. 4-4 shows a second-order modulator with feedback compensation paths and mixers in both filter stages [4.2]. The output of the first filter stage is upconverted before summing with the feedback signal. After summation, the input signal of the second filter stage is downconverted. These extra mixers can be omitted if the outer feedback path is upconverted only (Fig. 4-5).

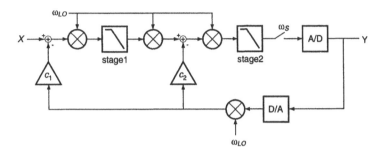

Fig. 4-4   *Multi-loop $\Sigma\Delta$ modulator with mixer inside the loop*

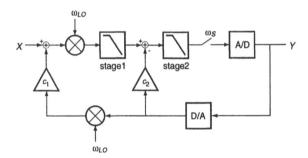

Fig. 4-5   *Multi-loop $\Sigma\Delta$ modulator with mixer inside the outer loop only*

4.2.2 Mixer outside the sigma-delta loop

A model of a continuous-time $\Sigma\Delta$ modulator with a mixer outside the sigma-delta loop is shown in Fig. 4-6. The IF input signal is downconverted to baseband by the mixer and is processed by a lowpass $\Sigma\Delta$ modulator. Placing the mixer outside the loop overcomes the problem of delay sensitivity in a continuous-time $\Sigma\Delta$ modulator. The main advantage of this topology in a continuous-time system is that the LO frequency and the sampling frequency do not have to be related and can be chosen independently. However, because the output of the mixer and the feedback path of the $\Sigma\Delta$ modulator are connected together, a fraction of the feedback signal may leak to the input of the mixer due to parasitic effects. As a result, a fraction of the high-frequency quantization noise around the mixer LO frequency may be mixed into the desired signal band. This effect is discussed in detail in section 4.3.3.

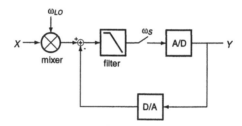

Fig. 4-6   *Model of a continuous-time $\Sigma\Delta$ modulator with mixer outside the loop*

4.2.3 Mixer inside loop versus outside loop

It has been shown that if a mixer is placed inside the $\Sigma\Delta$ loop, non-idealities of the mixer are suppressed by the loopgain in the case of a multi-bit feedback signal, or a $\Sigma\Delta$ modulator with an analog filter in the feedback path. Therefore, linearity requirements of the mixer are relaxed. It was shown in section 3.2.3 that this does not apply to a single-bit modulator. Also, if the mixer is outside the loop, non-idealities are not suppressed by the loop and higher linearity demands apply to the mixer. Moreover, a $\Sigma\Delta$ modulator with a mixer inside the loop requires accurate synchronization between the mixers in the forward and feedback paths. This is achieved almost perfectly in a discrete-time system. In a continuous-time modulator, a delay between the mixers may cause leakage of quantization noise around the LO frequency into baseband. This effect is reduced if the LO frequency is equal to (a multiple of) the sampling rate, because quantization noise is highly suppressed around multiples of the sampling frequency. A DAC feedback pulse with return-to-zero (RTZ) is insensitive to small delays, if the LO frequency is a rational fraction of the sampling frequency. In all above mentioned topologies, the sampling frequency and the LO frequency have to be related.

A mixer outside the loop avoids synchronization problems, and offers more freedom of choice for the mixer LO frequency, which does not have to be related to the sampling frequency. The major advantage is that the choice for the sampling rate can be optimized for dynamic range performance and power consumption of the $\Sigma\Delta$ modulator, without affecting the LO frequency. Moreover, we have chosen to use a one-bit quantizer in the $\Sigma\Delta$ modulator for its high linearity. Therefore, linearity of the mixer is not reduced when placed within the $\Sigma\Delta$ loop. For these reasons, the IF-to-baseband $\Sigma\Delta$ modulator topology with the mixer outside the loop is used for the design in this chapter.

4.3 IF mixer design

The design of the IF mixer is described in this section. An integrated design is shown of a passive mixer and an RC filter, which is optimized for high linearity performance, low-power consumption, and integration with the continuous-time loopfilter of a $\Sigma\Delta$ modulator. This mixer/filter implementation is used as the input stage for the IF-to-baseband $\Sigma\Delta$ modulator. Furthermore, aspects of linearity, mixer driving and self-mixing as a result of charge injection are discussed.

4.3.1 Mixer topology

The passive mixer topology of Fig. 4-7a, that is used for the design in this chapter, is built from resistors and transistor switches, performing the V/I conversion and frequency translation, respectively. The transistor switches periodically couple and cross-couple the input current signal. This way, the input signal with frequency ω_{in} is multiplied by the rectangular LO waveform with frequency ω_{LO}. The (downconverted) fundamental output frequency ω_{out} of the mixer equals

$$\omega_{out} = |\omega_{LO} - \omega_{in}| \qquad (4\text{-}1)$$

For proper operation, the passive mixer is connected to the virtual ground nodes of an amplifier. Due to the high gain and the feedback, the input nodes of the opamp are virtual ground nodes. Because the opamp input nodes are modulated by only a fraction of the input voltage, highly linear performance can be achieved. The conversion gain of this passive mixer is:

$$CG = \frac{\sin\pi\delta}{\pi\delta} \cdot \frac{R_L}{R_{in}} \qquad (4\text{-}2)$$

where δ is the duty cycle of the LO signal. The voltage amplifier of Fig.4-7a can be easily transformed into an RC filter, by replacing the feedback resistors with capacitors (Fig.4-7b). This topology is a very suitable implementation for the IF input stage of a continuous-time ΣΔ modulator with mixer [4.3]. The resistors R_{in} perform a linear V/I conversion of the input signal. The current signal is linearly modulated by the mixer switches. As it has no bias current, the passive mixer does not add to the power consumption of the modulator. The virtual ground nodes of the amplifier provide linear summation nodes for the input current and the DAC feedback current (R_{dac} in Fig.4-7b). Moreover, the virtual ground nodes offer good isolation of the LO signal and the feedback bitstream signal. Isolation is essential for an independent choice of the sampling frequency and LO frequency (section 4.3.3). In the next sections, the circuit of Fig.4-7b is analyzed in more detail, regarding linearity, LO driving, and charge injection.

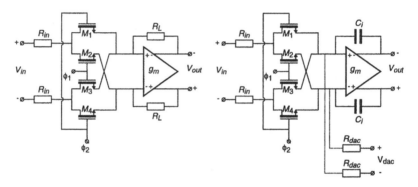

Fig. 4-7   *Integrated design of a passive mixer and voltage amplifier (a); passive mixer and active RC filter (b)*

4.3.2 Linearity performance

In this section, the non-linearity of the passive mixer topology of Fig.4-7b is analyzed. It has been assumed that the input (Poly-Si) resistors R_{in} are linear and that the mixer switches are connected to ideal virtual ground nodes (opamp has infinite gain). For distortion analysis, a mixer transistor is modeled as a non-linear impedance and an ideal switch (Fig.4-8) [4.3]. The switch impedance r_{on} is calculated by

$$r_{on} = \frac{\partial v_{ds}}{\partial i_{in}} \qquad (4\text{-}3)$$

where i_{in} is the current through the switch and v_{ds} the small-signal voltage across the switch. Using the model of a MOS transistor operating in the linear region, the switch impedance equals

$$r_{on} = \frac{L}{\mu_n C_{ox} W(V_{gt} - v_{ds})} \qquad (4\text{-}4)$$

where μ_n is the mobility, C_{ox} the oxide capacitance, W and L the transistor dimensions, and V_{gt} is the effective gate-source voltage obeying

$$V_{gs} = V_T + V_{gt} \qquad (4\text{-}5)$$

where V_T is the threshold voltage. In this analysis it is assumed that V_{gt} is constant. Having in mind that no bias current is flowing through the switch transistor (only signal current) the signal voltage v_{ds} across the switch is calculated using the small-signal parameter r_{on} of Eq. (4-4),

$$v_{ds} \approx \frac{r_{on}}{R_{in} + r_{on}} \cdot \frac{v_{in}}{2} \qquad (4\text{-}6)$$

where v_{in} was given in section 3.2.3. Likewise, the signal current through the input resistor and switch impedance can be calculated from Fig. 4-8

$$i_{in} \approx \frac{1}{R_{in} + r_{on}} \cdot \frac{v_{in}}{2} \qquad (4\text{-}7)$$

Substituting Eq. (4-4), Eq. (4-6), and Eq. (4-7) gives a non-linear expression for the signal current i_{in} (after Taylor series expansion)

$$i_{in} \approx \frac{v_{in}}{R_{in} + r_{on}} + \frac{3 \cdot r_{on}^3 \cdot v_{in}^3}{8 \cdot (R_{in} + r_{on})^4 \cdot V_{gt}^2} \qquad (4\text{-}8)$$

Using the result of Eq. (4-8), the expression for the third-order harmonic distortion (for a sinusoidal input signal) is approximately

$$HD3 \approx \frac{3}{32} \cdot \left(\frac{\hat{V}_{in}}{V_{gt}}\right)^2 \cdot \left(\frac{r_{on}}{R_{in}}\right)^3 \qquad (4\text{-}9)$$

assuming that the switch impedance is much smaller than the input resistance. Inspecting Eq. (4-9) reveals that distortion can be minimized by applying a large overdrive voltage V_{gt} to the mixer switches. Furthermore, the switch impedances should be made small compared to the input resistance. This is achieved by making the W over L ratio of the switch transistors large. The upper limit of the transistor size is determined by charge injection of the switches and speed requirements

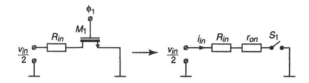

Fig. 4-8   *Mixer model (single-sided)*

Example 4.1: Consider the following data; V_{in} is 1 V (amplitude), V_{gt} is 250 mV, r_{on} is 1 kΩ, and R_{in} is 70 kΩ (v_{ds} is 14 mV). Substituting these parameters into Eq. (4-9) yields a HD3 of -107 dB. This example shows that high linearity of the mixer is achievable with this architecture.

4.3.3 Local oscillator driver

The mixer switches are connected to the input nodes of an opamp as well as to the feedback paths of the sigma-delta modulator (Fig. 4-7b). Being placed at the input of the sigma-delta modulator, any non-idealities of the mixer will have a direct impact on the modulator performance. This also applies to the local oscillator circuit that drives the mixer. The mixer has two clock inputs ϕ_1 and ϕ_2 with complementary phases. Ideally, the clock signals have exactly 180 degrees phase difference and infinitely steep rising and falling edges. However, due to parasitic effects it is not possible to generate these perfect complementary clock signals. Assuming infinitely steep edges, four different phases can be distinguished in case the inverter that is generating signal ϕ_2 has a small delay (Fig. 4-9). During phases φ_1 and φ_3, the clock signals are inverse to each other. During phases φ_2 and φ_4 the clock signals are both high and low, respectively. If the mixer is implemented with NMOS transistors, all switches are closed during phase φ_2 and open during phase φ_4. The dual situation occurs if the mixer is implemented with PMOS devices. When all switches are conducting at the same time, a low impedance path connects the feedback paths of the sigma-delta modulator as well as the input nodes of the opamp. In the following, two effects of the overlapping phases of the mixer clock are described. The first effect is due to the finite transconductance of the opamp and the second one is due to the offset.

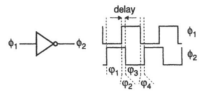

Fig. 4-9   *Overlapping mixer driver circuit (a); clock phases (b)*

Effect of finite transconductance

To analyze the effect of the overlapping clock phase φ_2 and the non-overlapping clock phase φ_4, the DAC current i_{dac}, flowing into the integration capacitors C_i, is calculated during all phases using Fig.4-7b. It has been assumed that the input voltage v_{in} is zero and that the amplifier does not have any offset. During phases φ_1 and φ_3 one pair of switches is open and one is closed. The current i_{dac} in phases φ_1 and φ_3 equals

$$i_{dac} = \frac{1}{2R_{dac} + \frac{1}{g_m} \cdot \left(1 + \frac{R_{dac}}{R_{in}}\right)} \cdot V_{dac} \qquad (4\text{-}10)$$

The right-hand term of the denominator can be neglected if the opamp has good virtual ground nodes (transconductance g_m is high) and if the DAC resistance R_{dac} and input resistance R_{in} have values that are in the same order (term between brackets in Eq. (4-10) is close to one). In phase φ_2 all switches are conducting and an equivalent resistance r_{on} (switch ON-impedance) is present between the input nodes of the amplifier. The current in phase φ_2 is

$$i_{dac} = \frac{1}{2R_{dac} + \frac{1}{g_m} \cdot \left(1 + \frac{2R_{dac}}{r_{on}}\right)} \cdot V_{dac} \qquad (4\text{-}11)$$

The term between the brackets is now much larger than one, if the DAC and input resistances are large and the switch impedance r_{on} small. Assuming that all mixer phases are within one DAC period (V_{dac} does not change), the difference between Eq. (4-10) and Eq. (4-11) represents the amplitude drop in DAC current during phase φ_2. Assuming that the input resistance is much larger than the switch impedance and that the DAC resistance is much larger than the inverse of the transconductance of the amplifier, the relative DAC current error is

$$\frac{\Delta i_{dac}}{I_{dac}} \approx \frac{1}{1 + g_m r_{on}} \qquad (4\text{-}12)$$

From Eq. (4-12) it can be seen that the larger the transconductance g_m, the smaller the amplitude drop. In other words, with stronger virtual ground nodes the amplifier is less sensitive to the low impedance of the mixer during phase φ_2. The relative charge error is calculated by

$$\frac{\Delta q_{dac}}{q_{dac}} \approx \frac{t_{\varphi 2} \cdot f_{LO}}{1 + g_m r_{on}} \qquad (4\text{-}13)$$

where $t_{\varphi 2}$ is the interval time of phase φ_2 and f_{LO} the local oscillator frequency. In phase φ_4 all switches are open and the input resistors are disconnected from the opamp input nodes. The current can be simply determined from Eq. (4-10) considering that the input resistors have become infinitely large

$$i_{dac} = \frac{1}{2R_{dac} + \frac{1}{g_m}} \cdot V_{dac} \qquad (4\text{-}14)$$

The same analysis as before can be done to calculate the relative charge error in phase φ_4. If $t_{\varphi 4}$ is the interval time of phase φ_4 the relative charge error is

$$\frac{\Delta q_{dac}}{q_{dac}} \approx \frac{t_{\varphi 4} \cdot f_{LO}}{2 g_m R_{in}} \qquad (4\text{-}15)$$

In section 4.3.2 it has been shown that an important design constraint for linearity of the mixer is that

$$R_{in} \gg r_{on} \qquad (4\text{-}16)$$

Comparing Eq. (4-13) and Eq. (4-15) shows that the charge error is much smaller in phase φ_4 than in phase φ_2.

Example 4.2: Consider the following data: g_m is 3.4 mA/V, R_{in} is 70 kΩ, r_{on} is 1 kΩ, $t_{\varphi 2} = t_{\varphi 4}$ is 200 ps, and f_{LO} is 10 MHz. From this data, it can be calculated that the relative charge errors are 0.045 % and 0.004 ‰ in phase φ_2 and phase φ_4 respectively. In this example the error in phase φ_4 is about 40 dB smaller than the error in phase φ_2.

With the data of the example above, two different clock schemes have been simulated. The first clock scheme has an overlapping and a non-overlapping phase (Fig.4-9) and the second scheme has two non-overlapping phases (Fig.4-21). We refer to the clock schemes as overlapping and non-overlapping, respectively. Furthermore, the sampling frequency of the $\Sigma\Delta$ modulator is 12.96 MHz and the local oscillator frequency is 6.48 MHz. The input signal is a 6.531840 MHz sinusoidal, which is mixed down to 51.84 kHz by the 6.48 MHz mixer. The (non)-overlapping phases are all 200 ps. For both simulations, the high-frequency quantization noise spectrum is similar and Fig.4-10 shows a detailed plot of this high-frequency quantization noise near the 6.48 MHz local oscillator carrier. Discrete tones are present at 6.368 MHz (-10.7 dB), 6.385 MHz (-10.8 dB), 6.420 MHz (-7.9 dB), 6.437 MHz (-7.9 dB), and 6.4717 MHz (-22.1 dB).

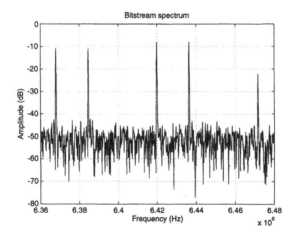

Fig. 4-10   *High-frequency quantization noise of IF $\Sigma\Delta$ modulator*

Fig.4-11 shows the simulated low-frequency output spectrum of the $\Sigma\Delta$ modulator with mixer that is driven by the overlapping clock scheme. The signal carrier is visible at 51.84 kHz and tones show up at 8.3 kHz (-92.9 dB), 43.5 kHz (-78.7 dB), 60.1 kHz (-78.6 dB), 95.4 kHz (-81.2 dB), and 112.0 kHz (-81.7 dB). The noise level gradually increases from about -120 dB to -110 dB (resolution bandwidth is 129.6 Hz). The in-band noise and frequency tones are mainly high-frequency quantization noise which is leaked in the signal band due to the overlapping (and non-overlapping) phases of the LO mixer driver.

The output spectrum of the ΣΔ modulator with mixer, driven by the non-overlapping clock scheme, is shown in Fig.4-12. Two small tones around the signal carrier can be observed at 44.5 kHz (-119.2 dB) and at 59.2 kHz (-118.8 dB). Again, these are high-frequency tones in the quantization noise spectrum at 6.48 MHz.

Consequently, in Fig.4-11 the high-frequency tones and noise leak into baseband with a gain of -70.8 dB, which agrees with the result of Eq.(4-13). The signal-to-noise ratio within the frequency band from 1 kHz to 100 kHz is 39.3 dB. The suppression of the tones in Fig.4-12 is simulated to be about 111 dB. The SNR is 58.5dB, which is close to the SNR in the case of ideal complementary clock phases (59dB). Compared to the simulation with overlapping clock signals, the resolution is improved more than 19 dB using a non-overlapping clock scheme.

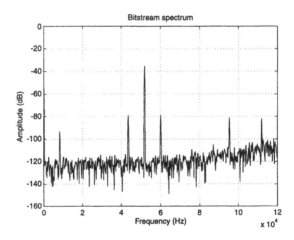

Fig. 4-11 *Simulated output spectrum of IF ΣΔ modulator with mixer driven by overlapping clock phases*

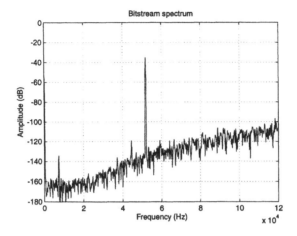

Fig. 4-12 *Simulated output spectrum of IF* $\Sigma\Delta$ *modulator with mixer driven by non-overlapping clock phases*

Effect of offset

In the following it has been assumed that the opamp has an offset voltage V_{os}, and the associated equivalent input referred offset current I_{os} is calculated in all phases. The offset current in phase φ_1 and φ_3 is the quotient of the offset voltage and parallel impedance of the input and DAC resistors

$$I_{os} \approx \frac{R_{in} + R_{dac}}{2 \cdot R_{in} \cdot R_{dac}} \cdot V_{os} \qquad (4\text{-}17)$$

Considering Eq. (4-16), the offset current in phase φ_2 approximately equals

$$I_{os} \approx \frac{1}{r_{on} - \frac{1}{g_m}} \cdot V_{os} \qquad (4\text{-}18)$$

Because of the low switch ON-impedance, the transconductance of the opamp can not be neglected. Inspecting the denominator of Eq. (4-18) learns that a large offset current is present during phase φ_2 as the offset voltage is divided by a small impedance. In phase φ_4 the input resistors are disconnected and the offset current is derived from Eq. (4-17) considering that the input resistors are infinitely large

$$I_{os} \approx \frac{1}{2R_{dac}} \cdot V_{os} \qquad (4\text{-}19)$$

which is much smaller than the offset in phase φ_2 because of the large DAC impedance.

> **Example 4.3:** Using the data of the previous example and if V_{os} is 3 mV and R_{dac} is 33 kΩ, the offset currents are calculated to be 4.25 µA and 45.5 nA in phase φ_2 and phase φ_4, respectively. The equivalent offset current can be calculated by multiplying Eq. (4-18) and Eq. (4-19) with the duty cycle of the overlapping and non-overlapping phase respectively.

4.3.4 Self-mixing

Clock feedthrough of the LO signal to the input of the mixer may result in an offset at the output of the mixer. This effect is referred to as self-mixing [4.4] and will be explained in this section. The gate capacitors of a MOSFET switch provide high frequency paths for the LO signal. The model of a switch including parasitic gate-capacitors is shown in Fig.4-13, where C_{gd} and C_{gs} are the gate-drain and gate-source capacitances respectively [4.5]. If the LO signal at the gates of the switches changes from low to high, or vice versa, charge is transferred to the input and output nodes of the mixer.

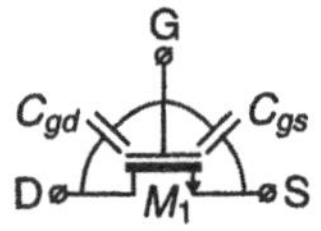

Fig. 4-13 *MOS transistor with parasitic gate-capacitors*

With the model of Fig.4-14 the self-mixing mechanism is explained. It is assumed that all switches are ideal (r_{on} is zero) and that perfect complementary LO clock signals are used. Only the parasitic gate-capacitors of switch S_1 have been included. In Fig.4-14a the switches S_1-S_2 are turned on and S_3-S_4 are turned off. The thick line shows the charge transportation path. The charge through both gate capacitors is pumped into the upper integration capacitor. The differential output voltage step as a result of charge injection of switch S_1 is

$$\Delta V_{out} = V_g \cdot \frac{C_{gs} + C_{gd}}{C_i} \qquad (4\text{-}20)$$

where V_g is the voltage swing at the gate of switch S_1. The other situation is shown in Fig.4-14b, where S_1-S_2 are switched off and S_3-S_4 are switched on. Now, the charge of the gate-source capacitor is pumped into the upper integration capacitor and the charge of the gate-drain capacitor is pumped into the lower integration capacitor through switch S_3. The output step equals

$$\Delta V_{out} = -V_g \cdot \frac{C_{gs} - C_{gd}}{C_i} \qquad (4\text{-}21)$$

Adding Eq. (4-20) and Eq. (4-21) gives the output voltage V_{out} after a mixing period, due to self-mixing

$$V_{out} = 2V_g \cdot \frac{C_{gd}}{C_i} \qquad (4\text{-}22)$$

Hence, the charge transfer through the gate-source capacitor is balanced out after one mixing period, while the charge of the gate-drain capacitor is modulated by switches S_1 and S_3, resulting in an equivalent offset current flowing into the integration capacitors.

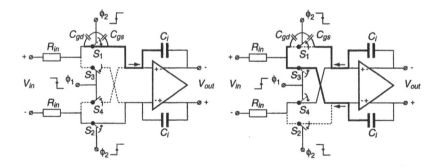

Fig. 4-14 *Model of mixer with parasitic capacitors of switch S_1. S_1 switching on (a); S_1 switching off (b)*

To minimize the effect of self-mixing, good matching of parasitic capacitors of the switches is required. If the switches have identical impedances and parasitic capacitors, all charge transfers are balanced and the offset is zero. To minimize charge injection, the switches should have small dimensions in order to have little parasitic capacitance. This design constraint is opposed to the linearity constraint, which requires switches with a large W/L ratio (section 4.3.2). To satisfy both constraints, the switches can be dimensioned best with a small channel length.

4.4 IF sigma-delta modulator design

In this section the design of a continuous-time $\Sigma\Delta$ modulator with mixer is described for A/D conversion of input signals, which are modulated at an intermediate frequency. The main target of this design is to show that highly linear A/D conversion of IF signals can be achieved at low-power consumption. The mixer is placed outside the $\Sigma\Delta$ loop. For high linearity of the DAC in the feedback path, a single-bit $\Sigma\Delta$ modulator is used.

4.4.1 IF sigma-delta modulator topology

The block diagram of the $\Sigma\Delta$ modulator with mixer is shown in Fig.4-15 [4.3]. The fully differential modulator includes a mixer at the input, a fourth-order continuous-time lowpass loopfilter [4.6] - [4.7], a single-bit A/D converter and a feedback path with D/A converter closing the $\Sigma\Delta$ loop. The input stage is the highly linear RC integrator with mixer topology of Fig.4-7b. The higher-order filter stages are transconductance-C integrators. The feedforward paths are implemented by means of transconductance amplifiers, which provide current output signals. Summing of the currents can be done by simply connecting all the output nodes. The comparator is designed to have current input capability. Each sampling period, the comparator takes a sample of the output signal of the loopfilter and determines the signal polarity. The output is a digital bitstream at the sample rate. The output data is fed back through a D/A converter which provides a positive or negative reference voltage. The reference voltage is converted into a current by the feedback resistors R_{dac} and subtracted from the input current. The filter stability coefficients have been determined according to the analysis in section 3.4.1. With a sampling rate of 13.0 MHz, which is a common GSM crystal oscillator frequency, the oversampling factor is 48 ($f_n = 270.833$ Hz).

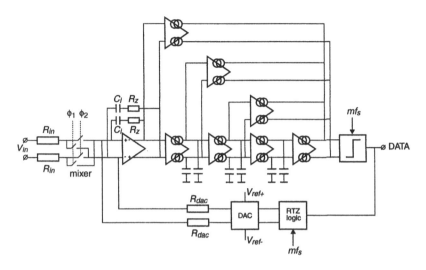

Fig. 4-15 *Block diagram of IF ΣΔ modulator, including mixer, anti-aliasing filter and A/D converter.*

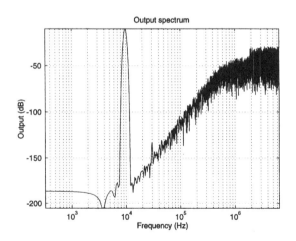

Fig. 4-16 *Simulated output spectrum of IF ΣΔ modulator with full scale 6.510 MHz input signal (LO is 6.5 MHz)*

With a fourth-order loopfilter, the (full signal) signal-to-quantization-noise in 100 kHz bandwidth is simulated to be 88 dB. Fig.4-16 shows the simulated output spectrum of the fourth-order $\Sigma\Delta$ modulator. In this simulation, the LO frequency is equal to half the sampling frequency (6.5 MHz) and the full-scale IF input signal is at 6.510 MHz. In the following, the designs of the individual blocks are described in more detail.

4.4.2 Input filter stage design

In Fig.4-17 the circuit design of the differential opamp with its biasing circuit is shown, that is used for the input filter stage. The opamp has a telescopic architecture implying that all transistors, including the common-mode transistors, are stacked in only two differential branches [4.8]. This way, current consumption of the opamp is a factor of two lower than of an opamp with a folded cascode architecture [4.9]. Due to the stacking of all transistors, the maximum output swing is limited. The differential input pair M_1-M_2, together with the cascode transistors M_3-M_4 boost the DC voltage gain of the differential input pair to a value of about 80 dB. The current sources M_7-M_8 also have cascode transistors M_5-M_6 in order to keep the output impedance high. Transistor M_9 is the PMOS current source of the opamp. Transistors M_{10}-M_{11} are biased in the linear region and the gates of these transistors control the common-mode output level, set by transistor M_{17}.

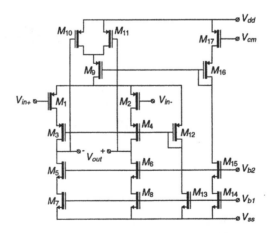

Fig. 4-17 *Differential inverting amplifier with biasing circuit*

The minimum supply voltage is determined by the saturation voltages of M_7, M_5, M_3, M_1 and M_9 (200mV each), the drain-source voltage of M_{10} (100 mV), and the maximum peak-to-peak swing at the output (400 mV$_{pp}$). Taking into account process spread, the minimum supply voltage is about 1.6 V. For proper biasing of the cascode transistors M_3-M_4 at full output swing, the output common-mode level is designed to be lower ($V_{dd}/3$) than the input common-mode level ($V_{dd}/2$).

Distortion versus supply current

In this section an expression for the low-frequency distortion of the RC integrator is derived. Distortion of the opamp of Fig. 4-17 is mainly caused by the non-linear differential input transistor pair M_1-M_2 [4.10]. Fig. 4-18 shows the non-linear model of the RC integrator in the $\Sigma\Delta$ loop. Due to the finite DC gain of the opamp, a small residue signal v_r is modulating the non-linear input pair. Using the square-law model of a MOS transistor [4.11], the residue voltage can be calculated (strong inversion)

$$v_r \approx \sqrt{\frac{2I_b}{\mu_p C_{ox}} \cdot \frac{L}{W}} \cdot \left(\sqrt{1 - \frac{i}{I_b}} - \sqrt{1 + \frac{i}{I_b}} \right) \qquad (4\text{-}23)$$

where i is the output current, I_b is the bias current of transistors M_1-M_2, L and W are the transistor dimensions, and μ_p the mobility. Taylor series expansion of Eq. (4-23) yields an expression for the linear and non-linear transconductance.

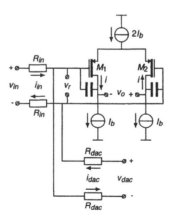

Fig. 4-18 *Model of input stage for distortion analysis*

Usually, in a differential system, the third-harmonic term is dominating the non-linear transfer

$$i \approx \frac{g_m}{2} v_r - \frac{g_m^3}{64 I_b^2} v_r^3 + \dots \tag{4-24}$$

where the transconductance g_m equals

$$g_m = \sqrt{2 \mu_p C_{ox} \frac{W}{L} I_b} = \frac{2 I_b}{V_{gt}} \tag{4-25}$$

where V_{gt} is the effective gate-source voltage. With the non-linear transconductance of Eq. (4-24) the relation between the linear residue current i_r and residue voltage v_r (Fig. 4-17) of the integrator is

$$i_r \approx \left(\frac{1}{R_{in}} + \frac{1}{R_{dac}} - g_m \right) \cdot v_r - \frac{g_m^3}{32 I_b^2} v_r^3 \tag{4-26}$$

where i_r equals

$$i_r = \frac{v_{in}}{R_{in}} + \frac{v_{dac}}{R_{dac}} \tag{4-27}$$

With Eq. (4-26), the expression for the residue voltage v_r is approximated as a function of the current i_r

$$v_r \approx \frac{1}{g_m'} \cdot i_r - \frac{g_m^3}{32 \cdot g_m'^4 \cdot I_b^2} i_r^3 \rightarrow g_m' = g_m - \left(\frac{1}{R_{in}} + \frac{1}{R_{dac}} \right) \tag{4-28}$$

The expression for the residue voltage is known, and the non-linear expression for the input current i_{in} and the feedback current i_{dac} can be extracted as a function of the residue current

$$i_{in} + i_{dac} \approx \frac{1}{2} \left(1 - \frac{1}{g_m'} \cdot \left(\frac{1}{R_{in}} + \frac{1}{R_{dac}} \right) \right) i_r - \left(\frac{1}{R_{in}} + \frac{1}{R_{dac}} \right) \frac{g_m^3}{64 g_m'^4 I_b^2} i_r^3 \tag{4-29}$$

Using the result of section 3.2.3, the relation for the third harmonic distortion (HD3) is determined from Eq. (4-29)

$$HD3 \approx \frac{\hat{V}_{in}^2}{64 g_m R_{in}^3 I_b^2}\left(1 + \frac{R_{in}}{R_{dac}}\right) = \frac{V_{gt}\hat{V}_{in}^2}{128(R_{in}I_b)^3}\left(1 + \frac{R_{in}}{R_{dac}}\right) \tag{4-30}$$

To minimize the bias current I_b, the input resistance R_{in} is designed to be as large as possible. The upper limit for R_{in} is set by the allowed thermal noise level (Eq. (4-31)). Moreover, the effective gate-source voltage V_{gt} of the input transistors should be made small (close to moderate inversion). This is done by designing the input transistors with a large W/L ratio. Table 4-1 shows the design parameters for the first integrator. Substituting these parameters into Eq. (4-30) gives an estimation of the third-harmonic distortion of the $\Sigma\Delta$ modulator, which is below -100dB.

Table 4-1   *Design parameters*

V_{dd}	2.5V
g_m	3.4mA/V
I_b	210μA
V_{in}	1.875V
R_{in}	70kΩ
R_{dac}	33kΩ

The input resistors and feedback resistors are dimensioned such that their noise contributions dominate quantization noise. The contribution of circuit noise of the opamp is neglected. In addition, due to the large signal bandwidth, thermal noise of the resistors dominates flicker noise of the opamp. The thermal noise of the resistors is

$$N_{th} = 8kTR_{in}\left(1 + \frac{R_{in}}{R_{dac}}\right)f_b \tag{4-31}$$

The resistor ratio R_{in}/R_{dac} is determined by the maximum input level and the DAC reference level according to

$$\frac{R_{in}}{R_{dac}} = \frac{1}{1 - RTZ}\cdot\frac{\sqrt{2}\hat{V}_{in}}{V_{dac}} \tag{4-32}$$

where RTZ is the return-to-zero time (relative to the sampling period). Minimizing the resistor ratio value implies a small RTZ interval and a large reference level V_{dac}. The upper limit of the input resistance R_{in} is determined by Eq. (4-31) and Eq. (4-32).

4.4.3 Transconductance-C integrator

The higher-order filter stages of the $\Sigma\Delta$ modulator are implemented as transconductance-C integrators. The integrators have voltage input and voltage output capability. Fig.4-19 shows the circuit design of the transconductance amplifier. The differential input pair M_1-M_2 convert the input voltage into a current. A degeneration resistor R_s between the sources of the input transistors increases the linear input range of the amplifier. Noise and distortion of the higher order integrators are not critical, as non-ideal behavior is suppressed by the gain of the previous stages. The output current flows into the integration capacitor between the output nodes of the amplifier. Because of the lack of feedback, the input impedance of the amplifier is very large. Transistors M_5-M_6 are NMOS current sources with cascodes M_3-M_4. The common-mode output level is controlled by transistors M_9-M_{10} as described in the previous section. The DC gain of all filter stages add to the total loopgain. Accordingly, the gain of each stage does not have to be high and 40 dB is sufficient. The input pair does not need cascode transistors for gain enhancement. The minimum supply voltage is equal to the saturation voltages of transistors M_5, M_3, M_1 and M_7 (200 mV each), the drain-source voltage of M_9 (100 mV), and the maximum peak-to-peak swing at the output and the maximum peak-to-peak swing at the source of M_1 (both 400 mV with 180° phase difference). Taking into account an extra 100 mV of spread, gives a minimum supply voltage of 2 V. The critical part in the design of the OTA is the biasing of the input transistors M_1-M_2. Due to the source degeneration, the full input swing at the gates of the input pair modulates the sources. As a number of transconductance-C integrators are cascaded, all integrators are biased at equal input and output common-mode levels ($V_{dd}/3$), as well as equal maximum input and output signal swings (800 mV$_{pp}$ differential).

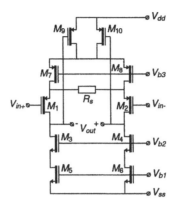

Fig. 4-19   *Circuit design of transconductance amplifier with source degeneration.*

4.4.4 A/D and D/A converter

The one-bit A/D converter design is shown in Fig. 4-20. It has a differential current input i_i which is converted into a voltage output by the latch M_1-M_2, with cascode transistors M_3-M_4 to increase latching speed. The latch is biased by current sources M_{11}-M_{12} with cascodes M_9-M_{10}. The output swing of the latch has been limited by diodes M_5-M_6, to ensure proper biasing of the NMOS current sources and cascode transistors at full output swing of the latch. The latch output signal V_o is fed into an inverter stage, producing a true digital output, which is stored into a flipflop. After the latch has made a decision, the latch output is reset to half the supply voltage by the CMOS switch M_7-M_8. If a mismatch exists between the capacitive loads of the differential output nodes of the latch, asymmetrical charge injection by the reset switch causes an offset. Due to this offset, the latch could be making incorrect decisions in the case of small input signals. However, these errors are not very critical as they are suppressed by the fourth-order loopfilter. The D/A converter consists of three switches, which connect the feedback resistors to the positive or negative reference, or a zero reference during the return-to-zero time. Because the feedback signal is a single-bit rectangular waveform, linearity of the DAC switches is not important and they can have small dimensions. Care is taken that both the positive and negative differential current pulses contain the same amount of charge.

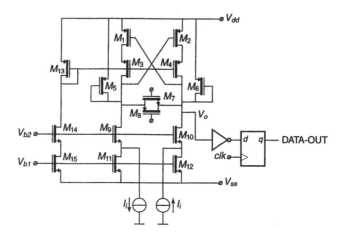

Fig. 4-20 *Single-bit A/D converter circuit*

4.4.5 LO driver scheme

The need for a non-overlapping mixer LO driver has been elaborated in section 4.3.3. A simple implementation of a circuit, generating non-overlapping clock phases, is shown in Fig.4-21a. It consists of two digital NAND ports (N_1 and N_2) two delay lines and two feedback paths. The NAND ports are driven by (non-ideal) complementary clock phases. Each NAND output is delayed by the delay line and fed back to the input of the complementary NAND port. The output buffers drive the mixer switches. The clock signals of the LO driver circuit are shown in Fig.4-21b. If the input CLK signal is low, the output node of N_1 is high. After some delay, the output of N_1 is fed back to the input of N_2. As both inputs of N_2 are high, the output of N_2 becomes low. If the CLK signal changes from low to high, the clock input of N_2 becomes low. Consequently, the output of N_2 is high. After some delay, both inputs of N_1 are high and its output changes from high to low. The inverting output buffer transform the overlapping clock phases into non-overlapping clock phases ϕ_1 and ϕ_2, which are never high at the same time.

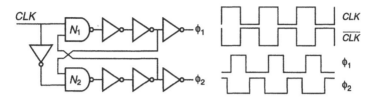

Fig. 4-21 *Non-overlapping LO mixer driver circuit (a); Clock phases (b).*

4.5 Experimental results

In this section the experimental results of two test chips are presented. The first test chip incorporates the fourth-order continuous-time $\Sigma\Delta$ modulator, without the IF mixer. The second test chip is the IF-to-baseband $\Sigma\Delta$ modulator of Fig. 4-15. The chip micrograph (active area) and partitioning of the $\Sigma\Delta$ modulator with mixer is shown in Fig. 4-22. The input stage and its biasing circuit roughly takes 50% of the total active area. The mixer area occupies only a very small part of the chip. The circuit has been designed in a 0.35 µm CMOS process with a single polysilicon layer, used for the resistors, and 5 metal layers. The active area of the test chip measures 0.2 mm^2. The LO driver circuit has not been integrated on chip for testing possibilities with different LO clock schemes.

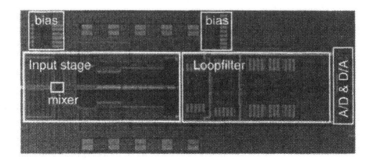

Fig. 4-22 *Test chip micrograph of IF $\Sigma\Delta$ modulator*

4.5.1 Test chip 1: Baseband sigma-delta modulator

Fig. 4-23 shows the measurement setup for the test chip. A Rohde&Schwarz audio analyzer provides a floating differential voltage signal which is connected to the input nodes of the chip. The bitstream output is connected to a 1-bit audio D/A converter which is synchronized with the master clock of the test chip. The analog DAC output is connected to the analog input of the Rohde&Schwarz analyzer, performing an 8192 points Fourier transform. Due to the 75 kHz bandwidth of the audio DAC, the end-of-the-band noise spectrum of the chip output signal is slightly suppressed. In all measurement plots, the maximum input level has been normalized to 0 dB. Signals and distortion tones, as well as noise are all relative to 0 dB. The analyzer bandwidth is 100 kHz, and the resolution bandwidth of the 8192 FFT is 12.2 Hz. All measurements have been done at a supply voltage of 2.5 V and a sampling rate of 13.0 MHz.

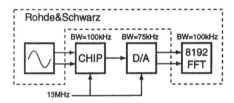

Fig. 4-23 *Measurement setup*

Distortion is measured with a full scale input signal. Fig. 4-24 shows the output spectrum of a 0 dB, 10 kHz input signal. The second and third-harmonic tones can be found at 20 kHz and 30 kHz respectively. Obviously, the second-harmonic is dominant at -90 dB, while the third-harmonic is well below -100 dB. It can be observed that the noise band around the 10 kHz carrier is slightly suppressed. Due to the limited dynamic range of the frequency analyzer, the signal carrier (and a small frequency band around it) has been filtered out, before calculating the FFT. A gain correction of the signal carrier is done afterwards. The noise reduction around the signal carrier has not been compensated and a small dip in the noise remains. Domination of the second-harmonic tone is not to be expected in a fully differential circuit. A possible cause may be an unbalance between the differential paths, which results in a finite suppression of even harmonic distortion in the differential circuit. The low third-harmonic tone shows the excellent linearity of the modulator which was to be expected from the linearity analysis in section 4.4.2.

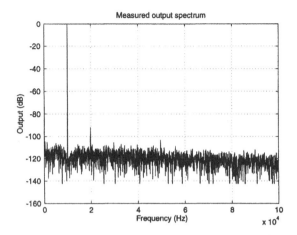

Fig. 4-24 *Measured output spectrum for a full scale (0 dB), 10 kHz input signal. The second-harmonic tone is at -90 dB, the third-harmonic tone is well below -100 dB.*

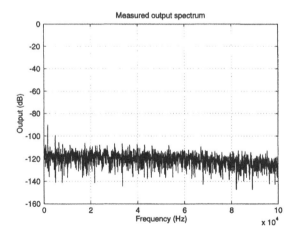

Fig. 4-25 *Measured output spectrum in the case of zero input signal. Spectral tones can be observed at 1.6 kHz and 4.8 kHz.*

Spectral tones are measured when no signal is applied to the input (section 3.2.2). The output spectrum is shown in Fig. 4-25. A tone of -90 dB is present at 1.6 kHz, and a smaller one at 4.8 kHz (-100 dB).

The signal-to-noise-and-distortion (SINAD) ratio has been measured as a function of the applied input level in Fig. 4-26. The input frequency is 10 kHz and the input level step size is 1 dB. From the SINAD plot, it can be seen that the curvature is almost flat up to the full scale input level. For small input signals the dynamic range is about 88 dB. For a maximum input level, the peak SINAD ratio drops to 86 dB, due to the second-harmonic distortion component and the increment of quantization noise. For signals larger than 0 dB the $\Sigma\Delta$ modulator is overloaded. Clipping of the integrators force the modulator into a stable second-order topology. If the modulator is in overload, large in-band distortion tones degrade the performance, and the SINAD ratio drops rapidly. At the end of the 100 kHz band, the quantization noise rolls off a little due to the filtering by the D/A converter in the measurement setup.

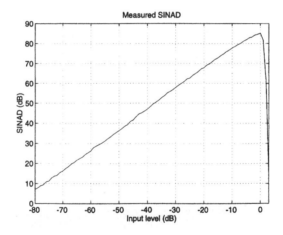

Fig. 4-26 *Measured SINAD ratio as function of applied input level (10 kHz input signal).*

4.5.2 Test chip 2: IF-to-baseband sigma-delta modulator

The second test chip incorporates the $\Sigma\Delta$ modulator and the mixer. An FM generator was used to provide the IF input signal. The gain of the passive mixer is -3.9 dB. In order to have the same maximum input level as the baseband test chip, the DAC reference voltage has been scaled down by 3.9 dB as well. The measured output spectrum is shown in Fig.4-27 for a -30 dBFS, 20 kHz sinusoidal input signal. The spurious tones and quantization noise are well below -100 dB. The SNR is measured to be 54 dB for the -30 dBFS input, resulting in a dynamic range of 84 dB. The DR is 4 dB lower than that of the first test chip, due to the downscaling of the DAC reference.

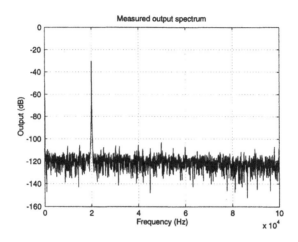

Fig. 4-27 *Measured output spectrum for a -30 dBFS, 13.020 MHz input signal (mixer LO frequency is 13.0 MHz).*

The next measurements show the effect of overlapping and non-overlapping mixer LO phases (Fig.4-28). The mixer is operating at half the sampling frequency (6.5 MHz) and a -30 dBFS, 6.549 MHz input signal has been applied. In the frequency band around this LO frequency, quantization noise power is high and the effect of overlapping mixer LO phases is large (dashed line in Fig.4-28). Clearly, strong tones are visible in the signal band and the noise level is slightly below -100 dB. The -30 dBFS signal carrier is at 49 kHz.

In the second measurement, non-overlapping clock phases have been used (solid line in Fig.4-28). The discrete peaks have disappeared and the noise level has dropped by 10dB to the same level as in Fig.4-27. No noticeable SNR degradation is measured if the non-overlapping clock phases are used.

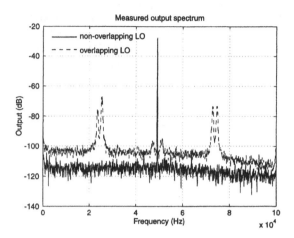

Fig. 4-28 *Measured output spectrum of -30 dBFS, 6.549 MHz input signal (mixer LO frequency is 6.5 MHz). The mixer is driven by overlapping LO phases (dashed) and non-overlapping LO phases (solid).*

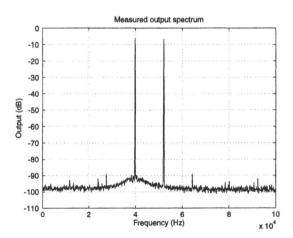

Fig. 4-29 *Output spectrum for IM3 measurement. Two -6 dBFS IF input tones at 13.040 MHz and 13.052 MHz are applied. The mixer LO frequency is 13.0 MHz*

Finally, linearity performance of the IF $\Sigma\Delta$ modulator is investigated. Two IF input signals of -6 dBFS at 13.040 MHz and 13.052 MHz have been applied and the mixer LO frequency is 13.0 MHz. As the two-tone signal generator already produces large inherent intermodulation distortion, two separate generators have been used, each generating a single tone. A linear signal combiner circuit has been designed which combines both tones and produces a differential two-tone IF output signal with low intermodulation distortion. Fig.4-29 shows the measured output spectrum. The average of 8 consecutive measured spectra has been calculated to get a clear view of the distortion tones. The -6 dBFS input tones are modulated by the mixer to 40 kHz and 52 kHz respectively. Third-order intermodulation distortion tones can be found at 28 kHz and 64 kHz respectively and are 84 dB down relative to the signal carriers. This results in a measured IM3 of -84 dB. The noise bulb around the 40 kHz carrier is phase noise from the 13.040 MHz signal generator. The generator producing the 13.052 MHz tone did have a clean spectrum around the carrier. Switching of the generator frequencies showed a clean spectrum around the 40 kHz carrier and phase noise around the 52 kHz carrier. The noise level of the spectrum of Fig.4-29 is higher than in Fig.4-27 (same resolution bandwidth). This observation is associated with the limited dynamic range of the measurement system, as has been the case for the distortion measurement of Fig.4-24. For the single tone measurement of Fig.4-24, this problem is solved by filtering out the signal carrier, before performing the FFT. The problem remains for the two-tone measurement of Fig.4-29, because only one of the tones could be filtered out by the frequency analyzer. Therefore, at a full scale two-tone measurement, dynamic range is limited by the analyzer.

4.5.3 Performance summary

The measured performance has been summarized in Table 4-2. Total power consumption of the $\Sigma\Delta$ modulator and mixer is 1.8 mW from a 2.5 V supply voltage. From 50 samples, 49 chips were tested to be functional. The sampling frequency of 13.0 MHz has been extracted from a GSM crystal oscillator. The oscillator and phase-locked-loop (PLL) have a total measured phase jitter of 4.5 ps.

Table 4-2   *Performance summary*

Parameter	Value	Unit
Technology	0.35μm CMOS (1P5M)	
Chip area	0.2	mm^2
Sampling rate	13.0	MHz
Bandwidth	100	kHz
Max. input	1.3	V$_{rms}$
Supply voltage	2.5	V
Power	1.8	mW
LO frequency	0-50	MHz
Dynamic range	82	dB
IM3 (f_{LO} = 13MHz)	-84	dB

4.6   Conclusions

In this chapter, the theory and design of a continuous-time ΣΔ modulator with integrated mixer have been described. Frequency translation can be performed inside or outside the ΣΔ loop. Placing mixers inside the loop for frequency translation, gives suppression of mixer non-idealities by the loopgain, if the ΣΔ modulator has a multi-bit feedback signal or a filter in the feedback path. Phase errors between mixers inside a continuous-time loop cause synchronization problems which result in quantization noise modulation with the mixer LO frequency. To overcome this problem the LO frequency and sampling frequency should be related. If frequency translation is done outside the ΣΔ loop, only a single mixer is required which does not have the problem of synchronization errors. As a result, the LO frequency and sampling frequency can be chosen independently.

It is shown that a passive mixer can be integrated with the RC integrator of the ΣΔ loopfilter for linear frequency translation. The RC filter has resistors for linear V/I conversion of the input and feedback voltage signals. The virtual ground nodes of an amplifier provide linear summation nodes of the signal currents. The passive mixer should be driven by a non-overlapping clock scheme to prevent modulation of the quantization noise spectrum with the LO frequency, by periodic shorting of the differential feedback paths.

Measurement results of a prototype test chip with the integrated design of a fourth-order continuous-time $\Sigma\Delta$ modulator and passive mixer topology show that high linearity of the mixer is achieved. Power consumption of the mixer is negligible. Driving the mixer with an overlapping clock scheme showed serious performance degradation for an LO frequency equal to half the sampling frequency. This effect of quantization noise modulation with the LO frequency is greatly suppressed if the mixer is driven with a non-overlapping clock scheme. The IF-to-baseband A/D converter realizes 82 dB of dynamic range in 100 kHz at a power consumption of only 1.8 mW. Measured third-order intermodulation distortion (IM3) for two -6 dBV input tones is -84 dB.

References

[4.1] Namdar A., and B.H. Leung, "A 400-MHz, 12 bit, 18-mW, IF Digitizer with Mixer Inside a Sigma-Delta Modulator Loop," *IEEE J. Solid-State Circuits*, pp. 1765-1776, Dec. 1999.

[4.2] Song, B.S., "A Fourth-order Bandpass Delta-Sigma Modulator with Reduced Number of Op Amps," *IEEE J. Solid-State Circuits*, pp. 1309-1315, Dec. 1995.

[4.3] Breems, L.J., E.J. van der Zwan, and J.H. Huijsing, "A 1.8-mW CMOS ΣΔ Modulator with Integrated Mixer for A/D Conversion of IF Signals," *IEEE J. Solid-State Circuits*, vol. pp. 468-475, Apr. 2000.

[4.4] Abidi, A.A., "Direct-conversion radio transceivers for digital communications," *IEEE J. Solid-State Circuits*, vol. 30, pp. 1399-1410, Dec. 1995.

[4.5] Gregorian, R., G.C. Temes, *Analog MOS integrated circuits for signal processing; design, manufacturing, and applications*, John Wiley & Sons, New York, 1986.

[4.6] Zwan, E.J. van der, "A 2.3 mW CMOS ΣΔ Modulator for Audio Applications," *ISSCC Dig. Tech. Papers*, pp. 220-221, Feb. 1997.

[4.7] Breems, L. J., E. J. van der Zwan and J. H. Huijsing, "A 1.8 mW CMOS ΣΔ Modulator for Mobile Communication," *Proc. of Workshop on Circuits, Systems and Signal Processing*, pp. 61-64, Nov. 1998.

[4.8] Nicollini, G., F. Moretti, M. Conti, "High-frequency fully differential filter using operational amplifiers without common-mode feedback," *IEEE J. Solid-State Circuits*, pp. 803-813, Jun. 1989.

[4.9] Huijsing, J.H., *Operational amplifiers; Theory and design*, Kluwer Academic Publishers, Boston, 2001.

[4.10] Breems, L. J., E. J. van der Zwan and J. H. Huijsing, "Design for Optimum Performance-to-Power Relation of a Continuous-Time ΣΔ Modulator," *Proc. of ESSCIRC*, pp. 318-321, Sep. 1999.

[4.11] Tsividis, Y.P., *Operation and modeling of the MOS transistor*, McGraw-Hill, New York, 1987.

Realization of a quadrature sigma-delta modulator

5

5.1 Introduction

In this chapter the design of a quadrature $\Sigma\Delta$ modulator with dynamic element matching circuit for image rejection improvement is presented. This A/D converter has been designed for application in an AM/FM radio receiver. In chapter 2 it was explained that image rejection performance in a receiver with frequency translation of the desired channel to a low-IF should be in the order of 80-90 dB. This amount of image rejection is achieved with a combination of channel selection and quadrature mixing. In this chapter, the image rejection performance of a quadrature IF-to-baseband $\Sigma\Delta$ modulator will be analyzed. In section 5.2 the main limitations of image rejection in the quadrature $\Sigma\Delta$ modulator are shown. A powerful way to improve matching and image rejection is DEM. In section 5.3.1 a DEM algorithm is presented which is controlled by the complex bitstream output of the quadrature $\Sigma\Delta$ modulator. The implementation of the DEM circuit is shown in section 5.3.2 and the effect of impedance mismatch and charge injection of the DAC and DEM switches are explained in section 5.3.3 and section 5.3.4 respectively. In section 5.4 the design of a prototype test chip is shown, and experimental results of the realized chip can be found in section 5.5. Finally, conclusions have been drawn in section 5.6.

5.2 Image interference

The design of the quadrature $\Sigma\Delta$ modulator in this chapter consists of two IF-to-baseband $\Sigma\Delta$ modulators as presented in chapter 4. The mixers of the modulators are driven by quadrature LO clock signals, to generate quadrature input signals with 90° phase difference. In section 2.4.3 it was explained that any gain or phase mismatch between the quadrature signals becomes evident in a finite image rejection ratio. In a $\Sigma\Delta$ modulator, non-idealities at the input of the system dominate the overall performance (section 3.2.3). Errors which are introduced further in the loop, are reduced by the gain of the preceding stages. This also applies to gain and phase mismatch between the I and Q signals in a quadrature $\Sigma\Delta$ modulator system. Fig. 5-1 shows a general model of a quadrature $\Sigma\Delta$ modulator. The input signal X is fed to the quadrature paths, which both incorporate a V/I converter and current mixer, producing quadrature signals X_I and X_Q. The quadrature loopfilter may consist of two real filters or a complex filter (section 3.5.2). Two sampling comparators generate the quadrature single-bit datastreams DATA-I and DATA-Q. The digital output is fed to the feedback loop that has a DAC and a V/I converter. Depending on the logic level of the DAC input signal, that can be a logic '1' or '0', the output of the DAC is either connected to the positive reference voltage ($+V_{ref}$) or negative reference voltage ($-V_{ref}$) respectively. The DAC output voltage is converted into a current by a V/I converter and subtracted as a current signal from the input current signal.

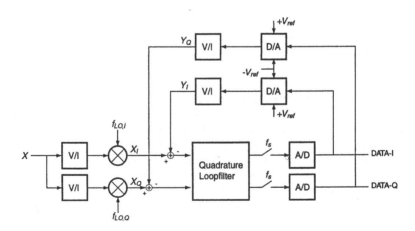

Fig. 5-1   *Model of a continuous-time quadrature $\Sigma\Delta$ modulator*

If it is assumed that ideal subtraction of the input and feedback signals is performed in the system of Fig. 5-1, the maximum image rejection ratio is limited mainly due to

- *Mismatch between D/A converters*
- *Mismatch between mixers and quadrature LO signals*
- *Mismatch between V/I converters*

In the implementation of the quadrature $\Sigma\Delta$ modulator of Fig. 5-1, the mixers and DACs consist of switches. Mismatch between the switches is negligible if the switch impedance is much smaller than the impedance of the V/I converters. Accurate phase matching between the quadrature LO signals is achieved by re-clocking these signals with a four times higher master clock [5.1]. The V/I converters have been implemented with polysilicon resistors. These resistors exhibit a parasitic capacitance to the underlying layer, and add a phase shift to the input and feedback signals. The V/I converters in the input paths are outside the $\Sigma\Delta$ loop, and a phase shift of the input signal does not affect stability. Therefore, the resistors can be designed with large dimensions (having large parasitic capacitance) for good matching. Moreover, as the signals through these V/I converters in the quadrature paths are identical, (matched) capacitive coupling of the signals from the I to the Q path (and vice versa) is not a problem. As a consequence, the resistors of both V/I converters can be placed in an interleaved configuration to improve matching. This does not apply to the V/I converters in the feedback paths. For stability of the loop, the parasitic delay of the resistors in the feedback circuit must be small. Therefore, small-sized resistors are required. Furthermore, parasitic coupling between the feedback paths adds to the total image interference of the quadrature modulator. Hence the feedback resistors of the I and Q paths should be placed in isolation from each other. Consequently, matching of the feedback resistors is less optimized than matching of the input resistors.

5.3   Dynamic element matching

To improve matching of the feedback resistors, dynamic element matching is used which is described in this section. Dynamic element matching is based on periodic exchanging of reference elements which may have a mismatch. As a result, the mismatch is modulated with the exchange rate to another (out-of-band) frequency. It was explained in the previous section that the V/I converters in the feedback paths are implemented with small-sized resistors that do not have very good matching. Therefore, in this section, different DEM algorithms are explored to improve the matching between the quadrature feedback paths of the $\Sigma\Delta$ modulator of Fig. 5-1. In all simulations in this section, a quadrature loopfilter with 5 complex integrators is used, having the same NTF as was shown in Fig. 3-34, with all notches being placed in the positive frequency band. The output spectrum of this modulator, in the case of ideal matching between the quadrature paths, was already shown in Fig. 3-36.

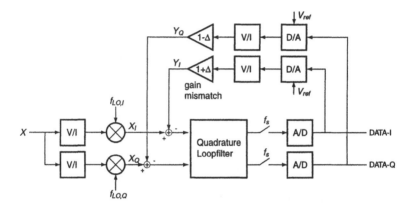

Fig. 5-2   *Quadrature $\Sigma\Delta$ modulator with gain mismatch between the quadrature feedback paths.*

A gain mismatch between the V/I converters (and the DACs) in the quadrature feedback paths can be modeled by adding an extra gain block in both paths (Fig. 5-2). It is assumed that, except for these gain blocks, the other functional blocks in the quadrature modulator are perfectly matched. Due to a gain mismatch, the amplitude of the analog feedback signal in the I-path Y_I is somewhat larger (or smaller) than the amplitude of the analog feedback signal in the Q-path Y_Q. In this section it is assumed that the normalized gain in the I-path is $1+\Delta$ and that normalized gain in the Q-path is $1-\Delta$, where Δ is the relative mismatch factor. In the time domain, the output signal of the quadrature modulator of Fig. 5-2 can be approximated by

$$Y \approx I + \Delta I + jQ - j\Delta Q \qquad (5\text{-}1)$$

where I and Q are the bitstream outputs of the I and Q modulators respectively, and the terms with Δ are the error signals. Rearranging Eq. (5-1) gives

$$Y \approx (I + jQ) + \Delta(I - jQ) \tag{5-2}$$

where the first term *(I+jQ)* is the (desired) output signal of a quadrature modulator (without mismatch) and the second term $\Delta(I\text{-}jQ)$ between brackets is the error signal *(E)* due to the gain mismatch Δ. Comparing both expressions in Eq. (5-2) reveals that the error signal of the Q modulator has an opposite sign in respect to its desired output signal. In the frequency domain this means that the (desired) output signal spectrum and the error spectrum (image) are mirrored around DC. In other words, a part of the negative frequency band 'leaks' into the positive frequency band (and vice versa). This is the essence of image interference.

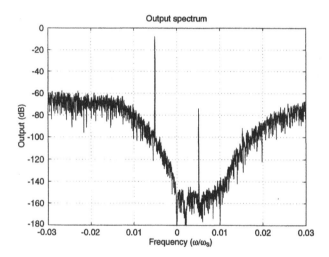

Fig. 5-3   *Simulated output spectrum of quadrature $\Sigma\Delta$ modulator with 0.1% gain mismatch between the feedback paths, without DEM.*

The effect of a gain mismatch between the feedback signals is simulated and the result is shown in Fig. 5-3. The desired signal band ranges from DC to $+0.01\omega_s$ rad/s, where ω_s is the sampling frequency, while the image band is in the negative frequency plane, ranging from DC to $-0.01\omega_s$ rad/s. An input signal is applied in the (negative) image band at $-0.005\omega_s$ rad/s. The quadrature feedback paths have a gain mismatch of 0.1% ($\Delta=0.05\%$). Due to the mismatch a signal occurs at $+0.005\omega_s$ rad/s. The simulated image rejection ratio is 66 dB (Fig. 5-3) which was expected from Eq. (2-4). Also, some quantization noise from the image band has leaked into the signal band (and vice versa) due to the finite amount of image suppression (section 3.5.2). Hence, mismatch between the feedback signals of a quadrature $\Sigma\Delta$ modulator causes leakage of noise and signals in the (negative) image band to the (positive) desired signal band (and vice versa).

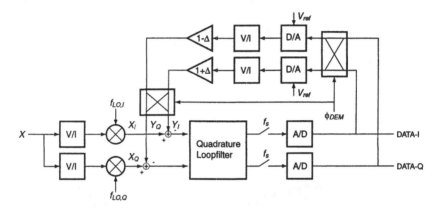

Fig. 5-4   *Model of a continuous-time quadrature $\Sigma\Delta$ modulator with DEM switches in the feedback paths*

Fig. 5-4 shows the model of the quadrature $\Sigma\Delta$ modulator with a DEM circuit. The DEM circuit consists of choppers at both sides of the quadrature feedback paths. The choppers consists of two pairs of switches that either couple or cross-couple the quadrature feedback paths. Thus, the feedback paths with the gain mismatch can be exchanged by the DEM choppers. If the DEM control signal ϕ_{DEM} is high (logic '1') the feedback paths are coupled straight through. If ϕ_{DEM} is low (logic '0') the feedback paths are cross-coupled. In this way, the gain mismatch is modulated by the DEM control signal ϕ_{DEM}, and the average gains of both DACs can be dynamically matched. A simple choice for the control signal of the DEM switches is a 50% duty cycle clock with a fixed frequency ω_{DEM}.

Fig. 5-5 shows the simulated output spectrum in the case of a 0.1% gain mismatch ($\Delta=0.05\%$) between the quadrature feedback signals and ω_{DEM} is $0.02\omega_s$ rad/s. As a result of the DEM, there is no 'leaked' image component present at $+0.005\omega_s$ rad/s (as was the case in Fig. 5-3). This image signal is modulated to spectral positions $+0.025\omega_s$ rad/s and $-0.015\omega_s$ rad/s, that are out of the signal band. Another important observation is the increased in-band noise level. This is due to the fact that the DEM circuit, besides modulation of the gain error, also performs demodulation of quantization noise present around (multiples) of the DEM frequency in the signal band. This amount of 'leaked' in-band noise can be substantial as the high frequency noise content of the feedback signals is large. A possible solution to reduce the amount of noise leakage, is to choose the DEM frequency equal to (a multiple of) the sampling frequency. This is due to the fact that around the sampling frequency, the quantization noise is greatly suppressed.

Pseudorandom selection of reference elements is another strategy for DEM [5.2]. When controlled by a pseudorandom signal, the gain mismatch between the DACs of Fig. 5-4 is randomized. However, the problem of in-band leakage of quantization noise remains with this DEM algorithm, as the quantization noise spectrum is modulated with a noisy pseudorandom spectrum. Fig. 5-6 shows the simulated result when the DEM circuit is controlled by a random signal at the sampling frequency. This means that at each sampling moment, the feedback paths are either coupled straight through or cross-coupled in a random order.

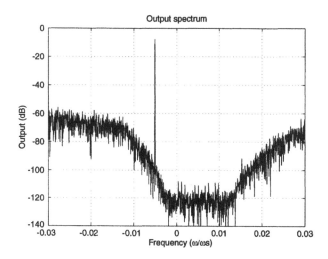

Fig. 5-5   *Simulated output spectrum of quadrature $\Sigma\Delta$ modulator with 0.1% gain mismatch between the feedback paths, ω_{DEM} is $\omega_s/50$*

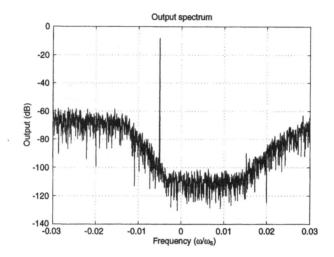

Fig. 5-6   *Simulated output spectrum of quadrature ΣΔ modulator with 0.1% gain mismatch between the feedback paths, with randomized DEM at ω_s.*

In the next section, another dynamic element matching technique is presented which is controlled by the complex bitstream output of the quadrature ΣΔ modulator. The matching algorithm applies to both continuous-time and discrete-time modulators, with lowpass or bandpass filters. The main advantage of this algorithm is that it does not have the problem of quantization noise demodulation, as is the case for the traditional techniques which have been discussed before. In section 5.3.1, the DEM algorithm is described in mathematical terms. The circuit topology is shown in section 5.3.2, and implementation related non-idealities, including impedance mismatch and charge injection of the DEM switches, are discussed in section 5.3.3 and section 5.3.4 respectively.

5.3.1 Complex data-controlled DEM algorithm

The essence of the complex data-controlled DEM algorithm in this chapter is that the mismatch error spectrum E in Eq. (5-2) is mirrored relative to DC, without causing demodulation of out-of-band quantization noise in the signal band. Mathematically, the following transformation is performed:

$$E \propto \Delta(I - jQ) \Rightarrow E^* \propto \Delta^*(I + jQ) \qquad (5\text{-}3)$$

where E^* is the transformation of the error signal E. In fact, the required transformation holds sign inversion of the error signal of the Q modulator (ΔQ). Replacing the error term in Eq. (5-2) by E^* yields

$$Y = (I + jQ) + \Delta^*(I + jQ) \qquad (5\text{-}4)$$

The signal spectrum and the transformed error signal spectrum are identical. Consequently, there is no leakage from the negative frequency band to the positive frequency band (and vice versa) and perfect image rejection is obtained. To accomplish the required transformation of Eq. (5-3) it is considered that I and Q are single-bit bitstreams and switching between (normalized) levels +1 and -1. An important property is that the square of these bitstreams is always +1

$$I^2 = Q^2 = 1 \qquad (5\text{-}5)$$

Using this result, the transformation as shown in Eq. (5-3) can be realized by modulation of the error signal E with the I and Q bitstreams [5.3]

$$E^* = I \cdot Q \cdot E = \Delta(I^2 \cdot Q - jQ^2 \cdot I) \qquad (5\text{-}6)$$

Using the results of Eq. (5-5), Eq. (5-6) simplifies into

$$E^* = \Delta(Q - jI) = -j\Delta(I + jQ) \qquad (5\text{-}7)$$

Indeed, Eq. (5-6) gives the required transformation of the error signal as defined in Eq. (5-3). It should be noted that, due to the transformation of Eq. (5-6), the phase of the gain error Δ has been shifted by 270° ($-j$). However, this does not affect image rejection as this phase shift is equal for both the I and Q modulators. In the next section, the implementation of the DEM algorithm of Eq. (5-6) is shown.

5.3.2 DEM implementation

In the previous section, a mathematical derivation was given of a dynamic element matching algorithm for the feedback paths of a quadrature $\Sigma\Delta$ modulator. The DEM algorithm is based on modulation of the mismatch between the feedback paths by the product of the digital outputs of the I and Q modulators (Eq. (5-6)). In this section the practical implementation of this algorithm is shown. Fig. 5-7a. shows an example of the bitstream signals I and Q in the time domain. At sample moment T_s (and $4T_s$) the bitstream signals I and Q are equal. Therefore, the product of the bitstreams $I \cdot Q$ is positive (+1) and, according to Eq. (5-6), the error E and the transformed error E^* have the same sign. At sample moment $2T_s$ (and $3T_s$) the bitstream signals I and Q are not equal. In this case the product of both bitstreams is negative and the error E and the transformed error E^* have opposite signs. Fig. 5-7b shows the logical representation of the algorithm, which is similar to an EXOR operation.

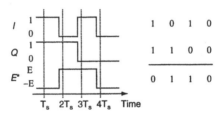

Fig. 5-7   *Transformation of error signal E; time domain (a); logic table (b)*

The quadrature $\Sigma\Delta$ modulator with the DEM circuit is shown in Fig. 5-8. The dynamic element matching is performed by mixers at both sides of the feedback paths. The DEM signal that drives the mixers is generated by a digital EXOR port. If the I and Q signals are not equal the output of the EXOR is high (logic '1'), and the feedback paths are coupled straight through. If the I and Q signals are equal, the output of the EXOR is low (logic '0'), and the feedback paths are cross-coupled. This way, the sign of the gain error in both quadrature paths is modulated according to the algorithm of Eq. (5-6). The system of Fig. 5-8 with the DEM circuit has been modeled and simulated. Again, the same loopfilter characteristic as in the previous simulations in this chapter has been used. In addition, a gain mismatch Δ of 0.5% is applied between the feedback paths. In Fig. 5-3, the simulated output spectrum was shown of this $\Sigma\Delta$ modulator, without DEM.

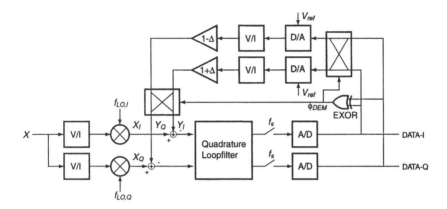

Fig. 5-8   *Model of a continuous-time quadrature ΣΔ modulator with complex data dependent DEM*

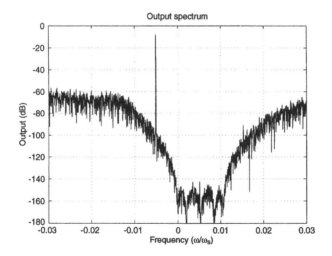

Fig. 5-9   *Simulated output spectrum of quadrature ΣΔ modulator with 0.1% gain mismatch between DACs, with DEM.*

The simulated output spectrum of the quadrature $\Sigma\Delta$ modulator with DEM is shown in Fig. 5-9. Obviously, in contrast to the output spectrum of Fig. 5-3, no image leakage is present. This shows that the DEM algorithm provides excellent gain mismatch correction. Moreover, no noise leakage from the negative frequency plane into the positive plane can be observed from Fig. 5-9. It should be noted that the same result is obtained if a phase mismatch were to be applied between the quadrature feedback paths.

Besides a gain (or phase) error, there may be a difference in the offset of both quadrature feedback paths. This occurs, for example, if the DACs in the quadrature feedback paths have different offset voltages. Fig. 5-10 shows the model of the quadrature $\Sigma\Delta$ modulator (with DEM circuit) where the DACs in the quadrature paths have offset voltages $+V_{os}/2$ and $-V_{os}/2$.

If the I and Q bitstreams are equal, the feedback paths are cross-coupled, while they are coupled straight through if both bitstreams are not equal. Consequently, the positive and negative offset voltages are added to the reference voltage of the I and Q modulators respectively in the case of unequal bitstreams, and added to the reference voltage of the Q and I modulators respectively in the case of equal bitstreams. Hence, the sign of the offset voltage is modulated according to the DEM algorithm, introducing an error

$$E \propto I \cdot Q \cdot V_{os} \qquad (5\text{-}8)$$

Eq. (5-8) shows that the offset voltage is modulated both by the I and Q bitstreams. This error has a noisy spectrum, as high frequency noise in the I bitstream is modulated with the high frequency noise in the Q bitstream. Therefore, due to the DEM circuit, an offset between the quadrature feedback paths causes in-band 'leakage' of high frequency quantization noise, which degrades SNR.

The effect of an offset in the quadrature $\Sigma\Delta$ modulator of Fig. 5-10 has been simulated with an offset voltage V_{os} equal to 0.1% of V_{ref}, where V_{ref} is the reference voltage of the DACs. The simulated spectrum is shown in Fig. 5-11a. The simulation plot shows an increased noise level and tones in the signal band. Fig. 5-11b shows the spectrum of I·Q and clearly the noise and tones of Fig. 5-11b are present in the spectrum of Fig. 5-11a, as was expected from Eq. (5-8). This simulation shows that, due to the DEM operation, different offset voltages in the quadrature feedback paths cause demodulation of high frequency quantization noise into the baseband.

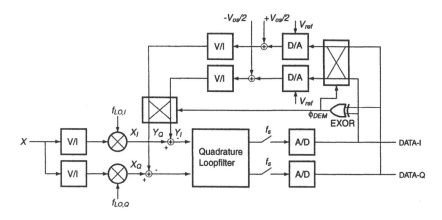

Fig. 5-10 *Quadrature* ΣΔ *modulator with offset voltage* V_{os} *between the feedback paths*

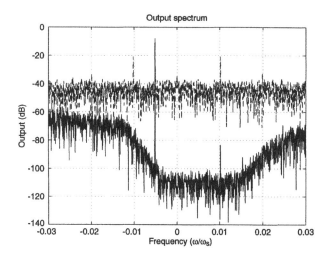

Fig. 5-11 *Simulated output spectrum of a quadrature* ΣΔ *modulator with an offset between the feedback paths (a, solid line); Spectrum of* $I \cdot Q$ *(b, dashed line)*

To overcome the problem of offset related quantization noise demodulation, a different feedback path topology than in Fig. 5-10 is used. Fig. 5-12 shows this improved model of the quadrature $\Sigma\Delta$ modulator with the DEM circuit. The positive reference voltage V_{ref} is converted into a current in both the I and Q feedback paths. Then, these reference currents are matched according to the DEM algorithm as described in this section. Finally, the bitstream related sign of the feedback current in each path is provided by means of a mixer, that is driven by the bitstream. If the data output of a modulator is high (logic '1'), the reference current has a positive sign, while it has a negative sign if the data output is low (logic '0'). In Fig. 5-12, the reference voltages in the I and Q feedback paths have been given different offset voltages. These offset voltages are being modulated by the DEM circuit as well as the mixers in both paths. The resulting error can be written as

$$E \approx I \cdot Q \cdot V_{os}(-I + j \cdot Q) = j(I + jQ)V_{os} \qquad (5\text{-}9)$$

The resulting error spectrum is similar to Eq. (5-7), and the offset only causes a slight gain and phase adjustment of the I and Q spectra. With the implementation of Fig. 5-12 mismatch in gain (and phase) and offset between the quadrature feedback paths of a quadrature $\Sigma\Delta$ modulator are matched by the DEM circuit. In the next section, the effect of mismatch between the switch impedances of the implemented DAC and the DEM circuits is explored.

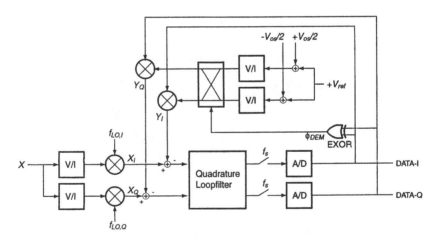

Fig. 5-12 *Improved model of quadrature $\Sigma\Delta$ modulator and DEM circuit.*

5.3.3 Impedance mismatch of DEM and DAC switches

In the design of chapter 4, the feedback of the $\Sigma\Delta$ loop incorporates a reference voltage, DAC switches, and resistors as V/I converters, which are connected to the input of the loopfilter. The error between the quadrature feedback signals in the $\Sigma\Delta$ modulator is mainly due to mismatch of the DAC resistors. Fig. 5-13 shows the differential implementation of the feedback circuit of Fig. 5-12. The DAC resistors are connected to the positive and negative reference voltages. The DEM switches interchange the DAC resistors. If the I and Q output bits are not equal, the resistors R_{dacI1} and R_{dacI2} are connected to the I-modulator, while resistors R_{dacQ1} and R_{dacQ2} are connected to the Q-modulator. If the I and Q output bits are equal, the resistor pairs are interchanged and R_{dacI1} and R_{dacI2} are connected to the Q-modulator, while R_{dacQ1} and R_{dacQ2} are connected to the I-modulator. Consequently, the mismatch between the DAC resistors is modulated according to the complex data dependent DEM algorithm of section 5.3.1. The DAC switches in the I and Q paths set the correct signs of the feedback signals.

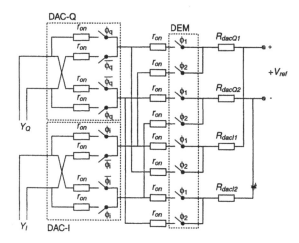

Fig. 5-13 *Implementation of DACs with DEM circuit (differential)*

To analyze the effect of impedance mismatch between the DAC switches, it is assumed that all switches of the DEM circuit are perfectly matched. Moreover, the DAC resistors are considered to be identical. Mismatch between the impedances of the DAC switches is not modulated by the DEM circuit and introduce a gain mismatch between the I and Q feedback signals. This gain mismatch is reduced by means of isolation and balancing. The amount of isolation is calculated as the ratio between the switch impedance and DAC resistance. The matching between the switches determines the balancing factor. If it is assumed that the switches in both DACs are perfectly matched, the gain mismatch Δa between the feedback signals is calculated by

$$\Delta a = \frac{r_{on}}{R_{dac}} \cdot \frac{\Delta r_{on}}{r_{on}} \tag{5-10}$$

where r_{on} is the average switch impedance and Δr_{on} is the mismatch between the switch impedances of the DACs in the I and Q paths. The first fraction of the right-hand side of Eq. (5-10) is the isolation factor and the second fraction is the balancing factor.

Example 5.1: Consider r_{on} is $1k\Omega$, R_{dac} is $100k\Omega$ and Δr_{on} is 100Ω, the gain error is 1E-3 (-60dB). With this gain error, the image rejection is 66dB.

Next, the effect of mismatch between the DEM switches is investigated. It has been assumed that all DAC switches and DAC resistors are perfectly matched. The first situation is that the switches in the DEM-I circuit in the I path are identical as well as the switches in the DEM-Q circuit in the Q path. A mismatch between the DEM-I and DEM-Q circuits (Fig. 5-14a) is similar to a mismatch between the DAC resistors and is compensated with the DEM algorithm. The second situation is that the set of DEM switches which are driven by signal ϕ_1 are matched, as well as the set of switches driven by ϕ_2. In the case of a mismatch between the two sets of switches (Fig. 5-14b), the I and Q path are always perfectly matched and no gain error is introduced. The third situation occurs if a mismatch is present between the switches of the DEM-I and DEM-Q circuits, as well as a mismatch between the switches driven by ϕ_1 and ϕ_2. Fig. 5-14c shows an example where there is a mismatch between the switches of DEM-I and DEM-Q, driven by ϕ_2, which also have a mismatch between the matched switches driven by ϕ_1. This way, an error is introduced which is modulated with the DEM algorithm. This error is demodulated by the I and Q bitstreams that are driving the DAC switches. The resulting error between the I and Q feedback signals is

$$E^* = \frac{\Delta}{2} \cdot (1 + I \cdot Q) \cdot (I - j \cdot Q) = \frac{\Delta}{2} \cdot (I - jQ) - j \cdot \frac{\Delta}{2} \cdot (I + jQ) \tag{5-11}$$

which consists of a gain adjustment of the signal and image spectra, as well as image leakage. In case of a mismatch error Δ of 0.1%, the image rejection is again 66 dB. In conclusion, impedance mismatch between the switches of the DAC and DEM circuits can result in image leakage, which can be minimized by means of isolation and balancing. Good isolation is obtained if the switch impedances are small compared to the DAC resistances, while balancing is optimized with the matching schemes of Fig. 5-14a or Fig. 5-14b.

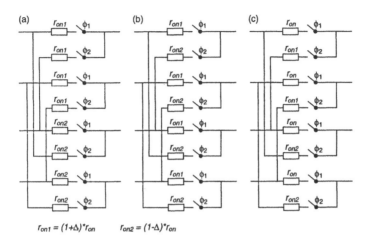

$$r_{on1} = (1+\Delta)^* r_{on} \qquad r_{on2} = (1-\Delta)^* r_{on}$$

Fig. 5-14 *Model of DEM switch impedances (normalized to 1), with mismatch between DEM-I and DEM-Q switches (a), between switches driven by ϕ_1 and ϕ_2 (b), between DEM-I and DEM-Q switches, and switches driven by ϕ_1 and ϕ_2 (c).*

5.3.4 Charge injection of DEM and DAC switches

The DAC switches as well as the DEM switches are connected to the virtual ground nodes of the opamp of the first filter stage. This implementation is, like the mixer (section 4.3.4) sensitive to asymmetric charge injection. First, the effect of charge injection of the DAC switches is analyzed. Fig. 5-15 shows the implementation of the DAC with a parasitic gate capacitor of switch S_1 included. The other switches are supposed to be ideal, without parasitics. For simplicity, only a single modul tor is taken into account without DEM switches. The switches are NMOS transistors, which are closed at a high gate voltage and open at a low gate voltage. Suppose, all switches are open initially (gates are low). If switches S_1-S_2 are closed and S_3-S_4 are opened, charge is injected through the gate capacitor of S_1 to the negative output node of the integrator. The output voltage is calculated by Eq. (4-20). If the switches S_1-S_2 are opened and S_3-S_4 are closed, charge is injected, with opposite sign, through the gate capacitor to the positive output node of the integrator. In both cases the differential offset at the output has the same amplitude and sign. This offset only occurs in the case that two consecutive output bits are not equal ('01' or '10' pattern). If the consecutive bits are equal ('00' or '11') the state of the DAC does not need to change and no switching is performed. Consequently, the gate capacitor of S_1 generates a small offset which is modulated with the exclusive-OR of two consecutive bits. In other words, the feedback signal is mixed with a one sample period delayed version of itself. This results in leakage of high frequency quantization noise into the signal band. This effect may be considered as 'self-mixing' of the DAC or a 'memory effect'. Suppression of this error is again obtained by isolation (small switch sizes) and balancing (matching of parasitic capacitors). However, these techniques may not provide satisfactory suppression. This is because the minimum switch size is determined by impedance requirements, and matching of parasitics may not be very reliable.

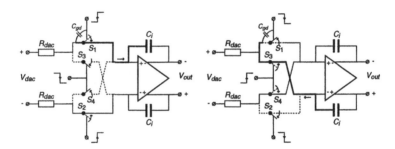

Fig. 5-15 *Implementation of the DAC with charge injection mechanism of parasitic gate capacitor of switch S_1*

To overcome the problem of self-mixing, the circuit of Fig. 5-16a is used. All capacitive loads at nodes 1 and 2 in Fig. 5-16 are modeled as capacitors C_{db1} and C_{db2}. The DAC switches S_1-S_4 are driven by the clock scheme of Fig. 5-16b. Each sampling period all DAC switches are open for a short time. During this 'off-line' time of the feedback, reset switches S_5-S_6 connect the input nodes of the DAC switches to a reference source (Fig. 5-16a). This way, the 'memory' charge which has built up in C_1 and C_2, flows into the reference node. The reset switches can have minimum size, as long as the switch impedances are small enough to ensure complete dischargement during the 'off-line' time. Charge injection of the reset switches S_5-S_6, when closed to clear the memory, is also discharged by the reference source. When the reset switches S_5-S_6 are opened after discharging, another charge error is dumped into the memory capacitors through charge injection of the reset switches S_5-S_6. This error can be small because of the minimum size reset switches. The main advantage is that this error is identical each sampling period and only causes a small gain error of the feedback signal. The clock phases of the DAC switches and reset switches should be non-overlapping to prevent a short-circuit path between the input nodes of the opamp. In order not to interrupt the feedback signal, the reset switches should be active during the RTZ interval of the DAC.

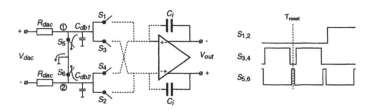

Fig. 5-16 *Implementation of the DAC with reset switches S_5-S_6 to clear the 'memory' capacitors (a); Clock scheme (b)*

So far, the DEM switches have not been taken into account. The DEM switches also suffer from charge injection and have the same problem of self-mixing. This problem is solved again with the reset switches as shown in Fig. 5-16. If the DEM switches are controlled during the reset time, charge that is injected by the DEM switches flows into the reference node. Fig. 5-17 shows the complete implementation of the quadrature DACs, the DEM circuit, and the reset switches, with the clock scheme. Initially, the I and Q bitstreams are 0 and 1 respectively. At sample moment T_s, the DAC switches in the I and Q paths are all switched open, and charge is injected into the capacitive loads of internal nodes 1-2 and 3-4. In the next sampling period, both bitstreams are 1 and the DEM circuit needs to exchange the DAC resistors. The DEM switches also dump charge into the capacitive loads of the internal nodes. The capacitive loads are discharged by the reset switches. After the reset interval, the reset switches are all opened and dump a small amount of charge into the 'empty' capacitive loads of the internal nodes. Each sampling period, this charge injection is identical and offset errors are introduced at the input of the DAC-I and DAC-Q switches. This offset is modulated with the I and Q bitstreams and only causes a small gain error between the quadrature feedback signals.

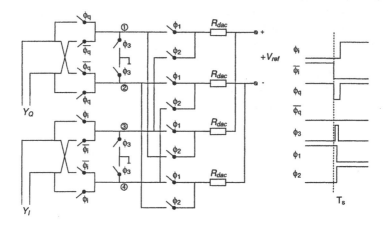

Fig. 5-17 *Implementation of DEM switches and DACs with reset switches (a); Example clock scheme (b).*

Another parasitic capacitor C_{ds}, between the drain and the source of the DAC switches, connects the input node of a DAC switch directly to the output. Fig. 5-18 shows the implementation of the DAC with two DEM switches S_7-S_8. The drain-source capacitance of S_1 and the parasitic gate capacitor of the DEM switch S_7 are included. For independent analysis of the effect of the drain-source capacitance, the total capacitive loads at nodes 1 and 2 are assumed to be equal and much larger than the drain-source capacitance. In the reset interval, switches S_1-S_4 are all open, while switches S_5-S_6 are closed. If the state of the switches S_7-S_8 changes during reset time, charge is injected via the gate-source capacitance of S_7 into node 1. Via the drain-source capacitance of S_1, charge is transported to the output of the integrator. The output voltage is calculated by

$$V_{out} = \frac{C_{gs}}{C_{db}} \cdot \frac{C_{ds}}{C_i} \cdot V_g \qquad (5\text{-}12)$$

Where V_g is the voltage swing at the gate of S_7. From Eq. (5-12) it can be learned that this is a second-order effect, as both fractions are much smaller than 1. In addition, this effect is compensated because nodes 1 and 2 are discharged by the reset switches and inject the same amount of charge with opposite sign.

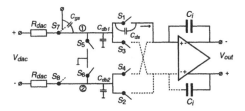

Fig. 5-18 *Charge transport mechanism via parasitic drain-source capacitor of switch S_1*

5.4 Quadrature sigma-delta modulator design

In this section the design of the quadrature $\Sigma\Delta$ modulator with the DEM circuit is described for quadrature mixing and A/D conversion of IF input signals. The main target of this design is to show that good matching of the quadrature feedback signals can be achieved using the data dependent DEM algorithm as presented in this chapter. The quadrature modulator consists of two IF $\Sigma\Delta$ modulators (chapter 4), both with a 5th-order loopfilter. The A/D converter is intended for low-IF conversion (section 2.2.1) in AM/FM radio receivers, with 300 kHz bandwidth.

5.4.1 Topology

The A/D converter includes two fifth-order continuous-time $\Sigma\Delta$ modulators, with passive mixers at the input. Fig. 5-19 shows the block diagram of one of the modulators. The mixers are driven by 10 MHz quadrature LO clocks with 90° phase difference. The sampling rate is 21.07 MHz, and the oversampling ratio of the modulator is 32. The fifth-order loopfilter has feedforward compensation paths for high-frequency stability of the loop. Two local feedback paths create resonator stages with complex conjugate poles. With these resonators, gain can be distributed equally within the signal band to minimize in-band quantization noise. The resonator notches are positioned at 165 kHz and 285 kHz respectively.

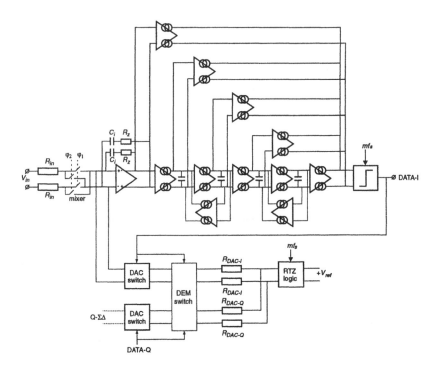

Fig. 5-19 *Block diagram of fifth-order $\Sigma\Delta$ modulator with DEM*

The simulated output spectrum of the quadrature $\Sigma\Delta$ modulator is shown in Fig. 5-20 (signal at 100 kHz). The design of the feedforward compensated lowpass filter with local feedback is described in section 3.4.2. Filter coefficients have been dimensioned such that stability of the loop is preserved at 20% spread in coefficient values. The reference voltage (2.8 V) of the DACs is generated by an on-chip bandgap reference. For the control of the DEM switches, a number of accurate clock phases are needed. A delay-locked loop (DLL) circuit has been implemented for generation of these clock phases. Jitter of the DLL is not critical, as the DEM circuit switches in the return-to-zero interval of the DAC pulse. Hence, timing accuracy of the DAC pulses is not affected. The designs of the mixer, the first integrator opamp, the feedforward coefficients and the comparator are similar to those as shown in chapter 4. The transconductance amplifiers in the higher-order stages and the local feedback paths are implemented by single-stage amplifiers with folded cascode transistors for high output impedance, and incorporate source degeneration for linearity. The quality factors of the resonators are limited by the finite output impedance of the transconductance amplifiers. In the following sections, special attention is paid to the design of the critical blocks of the modulator for image rejection, which are the LO driver circuit and the feedback circuit.

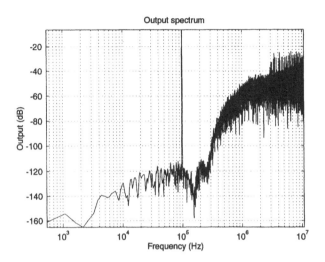

Fig. 5-20 *Simulated output spectrum of fifth-order quadrature $\Sigma\Delta$ modulator (positive frequency band)*

5.4.2 Input stage

The in-phase and quadrature-phase input current signals of the $\Sigma\Delta$ modulator are generated with the mixer circuit of Fig. 5-21. Two differential resistor pairs are used as V/I converters and the two passive mixers are driven by non-overlapping quadrature clock phases. The gain error between the input currents is determined by the mismatch of the input resistors, the mismatch of the quadrature clock phases, and the impedance mismatch of the mixer switches. To minimize the resistor mismatch, the resistors have been placed interleaved. Matching improves with the square root of the resistor area. As parasitic delay of the resistors is not critical, the resistor width has been dimensioned four times larger than minimum. The effect of mismatch between switch impedances can be calculated with Eq. (5-10). In the case of input resistances of 62.5 kΩ, and switch impedances of 1 kΩ with 100 Ω mismatch, the image rejection is 62 dB. The phase error is determined by the mismatch between the timing of the clock edges of the quadrature LO signals, and the mismatch between parasitic delay of the input resistors.

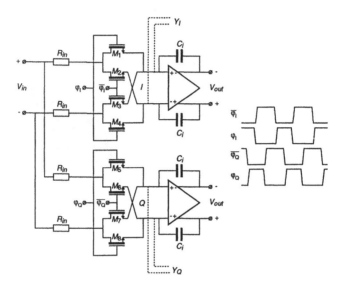

Fig. 5-21 *Quadrature mixer implementation (a); LO clock driver scheme (b)*

The quadrature LO clock signals are derived from a 40 MHz external master clock. The master clock frequency is divided by 2, to generate an accurate 50% duty cycle, 20 MHz clock. A second divider generates the 10 MHz LO frequency. Another 10 MHz clock with 90° phase difference is generated by means of an exclusive-OR function of 10 MHz clock and 20 MHz clock inputs. The 10 MHz quadrature LO clocks are fed into two flip-flop buffers, which are clocked with the 40 MHz master clock. This way, accurate phase matching between both LO clocks is obtained. Finally, two non-overlapping stages of section 4.4.5 generate non-overlapping quadrature LO clock phases to drive the IF mixers.

5.4.3 Feedback and DEM circuit

The complete implementation of the quadrature feedback circuit is shown in Fig. 5-22. All switches are NMOS transistors. The DACs in the I-path and Q-path are implemented by switches M_1-M_4 and M_5-M_8 respectively. Switches M_9-M_{10} connect the reference voltages to the feedback circuit, while switches M_{28}-M_{29}.

connect the feedback circuit to a zero reference voltage during the return-to-zero interval. Switches M_{11}-M_{14} are the reset switches, which clear the capacitive loads of the D/A converters, by connecting the drains of the DAC switches to the zero reference source. Switches M_{15}-M_{22} form the DEM circuit. The DEM switches are controlled during the reset interval to shield non-ideal effects of the switching from the outputs of the DACs. The RTZ switches and the reset switches can be minimum-sized transistors.

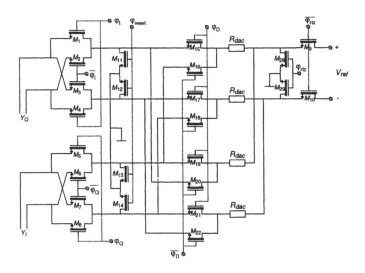

Fig. 5-22 *Implementation of feedback paths with DACs and DEM circuit*

For the purpose of linearity and isolation, the DAC and DEM transistors have been dimensioned with ON-impedances which are about a factor 100 smaller than the DAC resistances R_{dac} (140 kΩ). It was already shown in Fig. 5-17b that a lot of accurate clock phases are required for driving of the switches in the feedback circuit. These phases have been generated with a delay-locked loop circuit. Basically a DLL consists of a delay line with N delay cells and a phase detector. The phase detector of the DLL detects the phase difference between the input signal and output signal of the delay line. Depending on the phase difference, the delay time of the delay cells is increased or decreased by adjusting the bias current of each delay cell. This way, the phase difference between the input and output signals of the delay line is controlled to zero, with exactly one period delay between the signals. Consequently, the phase shift of each delay cell is accurately controlled to $1/N^{th}$ of the sampling period and N different clock phases are available from the delay line. The delay of each cell is insensitive to temperature, process spread, and the supply voltage. With the clock phases from the delay cells, the clock scheme of Fig. 5-23 can be derived. With the clock scheme of Fig. 5-23 the strategy of switching is explained.

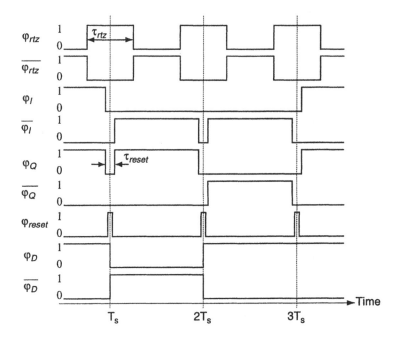

Fig. 5-23 *Clock scheme of switches in feedback circuit*

The feedback signals incorporate a RTZ time of half the sampling period. During the RTZ interval τ_{rtz}, the DAC switches are all opened for a short (reset) time τ_{reset}, and disconnect the feedback circuit from the input of the quadrature modulator. During the reset time, the reset switches are closed and connect the input nodes of the DAC switches to the zero reference source. Also, in the reset interval the DEM switches are controlled, if necessary. At the end of the reset interval, the input nodes of the DAC switches are discharged and the reset switches are opened. Finally, the feedback loop is closed again with the DAC switches. After the RTZ interval the reference voltages are switched on and new feedback pulses are applied to the input filter stages of the quadrature modulator.

5.5 Experimental results

In this section, the experimental results are given of a test chip (Fig. 5-24) for AM/FM radio which incorporates the complex $\Sigma\Delta$ modulator and the DEM circuit as described in the previous section. The circuit has been designed in a 0.35 μm CMOS process using a single polysilicon layer and 5 metal layers. The DEM circuit only occupies a small part of the total chip area. The sampling frequency has been derived from a 42.1 MHz crystal oscillator and is 21.05 MHz. The LO frequency of the mixers is 10 MHz. All measurements have been done at a 3.3 V supply. The IF input signal at 10.1 MHz is generated by a Marconi multi-tone generator. The quadrature bitstream outputs (I and Q) are retrieved by an HP 1673G logic analyzer. The output file with the I and Q data is processed with MATLAB (complex adding and FFT). The signal bandwidth of the complex $\Sigma\Delta$ modulator is 300 kHz. In all measurement plots, the maximum input level has been normalized to 0 dB. The resolution bandwidth of the measured spectra is 40 Hz. Evaluation of the DEM algorithm involves two observations, which are the image rejection ratio and the in-band noise level. For the testing of the DEM circuit, the feedback resistors in the quadrature paths have been designed with a 20% mismatch. With this large mismatch, any increase of noise within the signal band, as a result of the DEM algorithm, can be easily observed. Fig. 5-25 shows the measured complex output spectrum of the quadrature $\Sigma\Delta$ modulator without the DEM circuit operating. The signal tone is +100 kHz, while the image component is at -100 kHz. The measured image rejection is 20 dB which agrees with a 20% mismatch. The small noise band around the +100 kHz signal carrier is phase noise from the external LO clock generator.

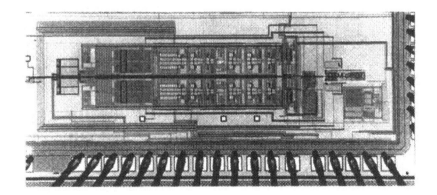

Fig. 5-24 *Test chip micrograph*

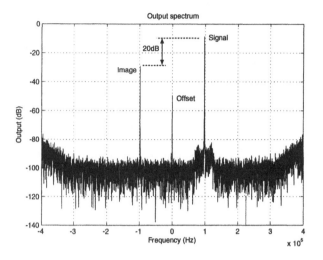

Fig. 5-25 *Measured spectrum without DEM*

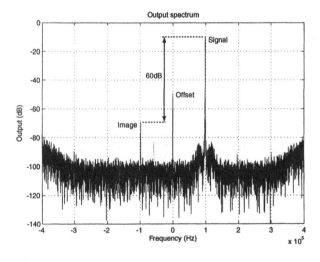

Fig. 5-26 *Measured spectrum with DEM active*

In the next measurement, the DEM circuit is active and the feedback resistors with the 20% mismatch are dynamically matched. Fig. 5-26 shows the measured output spectrum. The same input signal is applied as in the previous measurement. Clearly, the image component has dropped down considerably. The measured image rejection ratio is 60 dB, which is an improvement of 40 dB. Moreover, compared to Fig. 5-25, no increased noise level can be observed in the signal band when the DEM circuit is active. The image rejection ratio is limited by the mismatch in other parts of the circuit such as the input resistors, mixers, LO signals, and the DEM and DAC switches.

Next, measurements have been done with two FM input signals at 10.1 Mhz (image channel) and 9.9 MHz (desired channel) with 100 kHz and 25 kHz bandwidths respectively. The maximum levels of the downconverted image (at + 100 kHz) and desired (at -100 kHz) channels are -25 dB and -60 dB respectively. Fig. 5-27 shows the measured spectrum without DEM (20% initial mismatch). The desired channel at -100 kHz is not visible as it is concealed completely by the large -45 dB image leakage signal. Fig. 5-28 shows the same measurement, now with the DEM circuit operating. The leaked image component has dropped more than 40 dB and the desired channel at -100 kHz can be detected clearly. This measurement shows that the DEM circuit is also working for multi-tone FM input signals.

The IRRs of a batch of 37 test chips have been measured and the distribution of the measured results are shown in Fig. 5-29. All measured chips show an IRR better than 55 dB with DEM. From the figure it can be observed that the typical IRR is in the order of 60 dB with the DEM circuit. The straight line in Fig. 5-29 is an ideal normal distribution. The measured values closely match the ideal normal distribution with a likelihood of 98.8%. In case of a normal distribution, a mean (μ) and sigma (σ) value can be calculated which are 63 dB and 4.5 dB for these measured values.

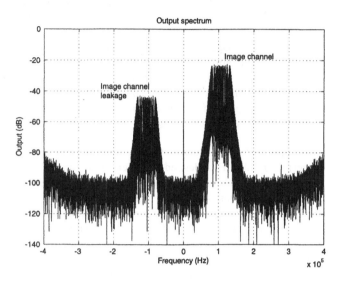

Fig. 5-27 *Measured spectrum with FM input without DEM*

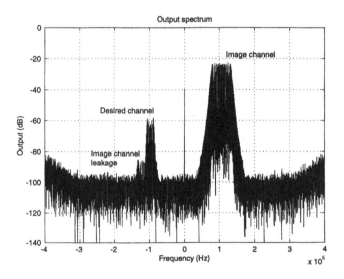

Fig. 5-28 *Measured spectrum with FM input with DEM*

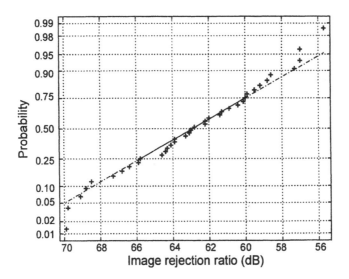

Fig. 5-29 *Distribution of measured data (37 samples)*

5.6 Conclusions

In this chapter, a DEM algorithm is presented that uses the quadrature bitstream outputs of a complex $\Sigma\Delta$ modulator. According to this DEM algorithm, the quadrature feedback paths of the $\Sigma\Delta$ modulator can be dynamically matched without causing demodulation of high frequency quantization noise, as is the case with traditional DEM techniques (fixed frequency or pseudo-random). The DEM signal is generated using an EXOR port that calculates the product of the I and Q bitstream signals. With this DEM algorithm, the error spectrum due to a mismatch between the quadrature feedback paths, is mirrored relative to DC.

A 0.35 µm CMOS prototype test chip with DEM circuit has been designed. A large systematic mismatch of 20% has been applied between the I and Q feedback resistors. Without DEM, the measured IR ratio is 20 dB, which agrees with the 20% mismatch. With DEM, the measured IR ratio is better than 55 dB for 37 different samples. No increased in-band noise level is observed, even with the large initial 20% mismatch. The prototype test chip was designed to test the DEM algorithm in the feedback path only, and no compensation techniques have been applied to improve the matching between the input circuits of the quadrature modulator (input resistors, mixers and LO drivers). Therefore, the image rejection is limited by the matching between these circuits to a typical value of 63 dB for 37 measured samples.

References

[5.1] Zwan, E.J. van der, K. Philips, C. Bastiaanse, "A 10.7 MHz IF-to-Baseband $\Sigma\Delta$ A/D Conversion System for AM/FM Radio Receivers," *ISSCC Dig. Tech. Papers*, pp. 340-341, Feb. 2000.

[5.2] Carley, L.R., "A noise-shaping coder topology for 15+ bit converters," *IEEE J. Solid-State Circuits*, vol. 24, pp. 267-273, Apr. 1989.

[5.3] Breems, L.J., E.C. Dijkmans, J.H. Huijsing, "A Quadrature Data-Dependent DEM Algorithm to Improve Image Rejection of a Complex $\Sigma\Delta$ Modulator," *ISSCC Dig. Tech. Papers*, pp. 48-49, Feb. 2001.

Benchmark

6

6.1 Introduction

In this book, some designs of low-power, high-performance IF-to-baseband A/D converters based on the principle of ΣΔ modulation are described. Optimizing the performance-to-power ratio, which is expressed by the FOMs in chapter 2, is done at three levels in the design process. In chapter 2, some system architectures are discussed which put different requirements on the A/D converter. The approach of this book is that IF-to-baseband A/D conversion can be performed with low-power and high-performance by combining a mixer with a continuous-time lowpass ΣΔ modulator. At the circuit level, an integrated filter and passive mixer design is proposed in chapter 4 that provides highly linear mixing at the cost of no extra power. Finally, at the transistor level, relations between distortion and design parameters have been derived to optimize the performance-to-power ratio. As mixing is done in the analog domain, good matching is required between the quadrature paths of the IF-to-baseband A/D converter for a sufficient amount of image rejection. Therefore, in chapter 5, a dynamic element matching algorithm is presented that improves matching between the quadrature feedback paths of the IF-to-baseband ΣΔ modulator. In chapters 4 and 5, the designs and experimental results of some test chips are presented. Special attention has been paid to the linearity of the modulators and matching performance of the quadrature paths.

6.2 Benchmark of test chips

To put the results of the test chips in the book in perspective, the measured performance is compared with that of other designs found in literature. First, the performance of the lowpass $\Sigma\Delta$ modulator of section 4.5.1 is investigated. For comparison of the different sigma-delta modulators the figure-of-merit of Eq. (2-6) has been used. The results of the benchmark are listed in Table 6-1. The top reference is the test chip of section 4.5.1 and the other references are all sigma-delta modulators from literature. Other types of A/D converters are not included in the list. For some papers, the SNDR has been determined from the measurement plots, in case the number was not published. Moreover, only the power consumption of the sigma-delta modulator has been taken into account (power of digital decimation filters not included). The different modulator types are continuous-time (CT), switched-capacitor (SC), and switched-opamp (SO).

Table 6-1   *Overview of baseband $\Sigma\Delta$ modulators*

Ref.	Type	BW	SNDR	Power	FOM (10^{-4})
design 1	CT, 4	100kHz	86dB	1.8mW	3.7
[6.1]	SC, 4	1MHz	85dB	230mW	0.2
[6.2]	SC, 2	200kHz	82dB	10.2mW	0.5
[6.3]	SC, 2	20kHz	91dB	67.5mW	0.06
[6.4]	SC, 2	20kHz	86dB	13mW	0.1
[6.5]	CT, 4	3.4kHz	74dB	0.2mW	0.07
[6.6]	SO, 3	16kHz	62dB	40µW	0.1
[6.7]	SC, 4	20kHz	88dB	1mW	2.1
[6.8]	SC, 3	1.25MHz	89dB	295mW	0.6
[6.9]	SC, 4	1.25MHz	88dB	105mW	1.2
[6.10]	SC, 4	1.1MHz	82dB	200mW	0.1
[6.11]	SC	25kHz	95dB	2.5mW	5.2
[6.12]	SC, 3	25kHz	98dB	47mW	0.6
[6.13]	SC, 3	160kHz	93dB	65mW	0.8
[6.14]	CT, 4	20kHz	94dB	2.3mW	3.6

Table 6-1   *Overview of baseband ΣΔ modulators*

Ref.	Type	BW	SNDR	Power	FOM (10^{-4})
[6.15]	SC, 2	3.2kHz	80dB	2.0mW	0.03
[6.16]	SC, 2-1	25kHz	80dB	5.4mW	0.08

From Table 6-1, the plot of Fig.6-1 is derived which shows the FOM as a function of the signal bandwidth. Clearly, the performance of test chip 1 is comparable with that of high quality audio ΣΔ modulators from literature.

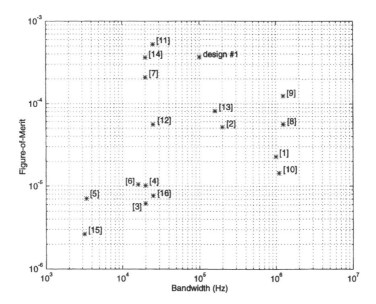

Fig. 6-1   *Figure-of-merit of lowpass ΣΔ modulators of Table 6-1 as a function of signal bandwidth*

The experimental results of the second test chip (section 4.5.2) are compared with that of other IF-to-baseband ΣΔ modulator designs. Also bandpass ΣΔ modulators are included in the benchmark, as they perform the same task (only with digital IF-to-baseband frequency translation). As linearity of these IF A/D converters is expressed by the intermodulation distortion rather than SNDR, an extra column for the measured IP3 is added in Table 6-2. This number is not published in all references, and for some designs, the IP3 is calculated from the measured intermodulation distortion. The different types are lowpass $\Sigma\Delta$ modulators with mixer (LPM), bandpass modulators (BP), and bandpass modulators with mixer (BPM). The fourth column of Table 6-2 shows the input IF. To be able to compare all designs, the FOM is based on the DR (Eq. (2-5)).

Table 6-2   *Overview of $\Sigma\Delta$ modulators for IF A/D conversion*

Ref.	Type	BW	IF	DR	IP3	Power	FOM (10^{-5})
design 2	LPM.	100kHz	13MHz	82dB	+36dBV	1.8mW	15
[6.17]	LPM	20kHz	10MHz	78dB	+18dBV	0.25mW	8
[6.18]	BP	100kHz	3.75MHz	71dB		130mW	0.02
[6.18]	BP	200kHz	3.75MHz	67dB		130mW	0.01
[6.19]	BP	200kHz	10.7MHz	57dB	+28dB	60mW	0.003
[6.20]	BP	30kHz	2MHz	56dB		0.8mW	0.03
[6.21]	BP	200kHz	20MHz	75dB		49mW	0.2
[6.22]	BPM	200kHz	81MHz	72dB	+6dBV	10.2mW	0.5
[6.23]	BP	200kHz	9.15MHz	72dB		60mW	0.09
[6.24]	BPM	200kHz	100MHz	49dB		330mW	0.0001
[6.25]	LPM	40kHz	400MHz	72dB		18mW	0.06
[6.26]	BP	2MHz	16MHz	75dB		140mW	0.7
[6.27]	BP	200kHz	200MHz	68dB		64mW	0.03
[6.28]	BP	200kHz	10.7MHz	78dB		80mW	0.3

Fig.6-2 shows the FOMs of the IF-to-baseband $\Sigma\Delta$ modulators of Table 6-2 as a function of the intermediate frequency of the input signal. The FOM of test chip 2 is more than a factor of 20 larger than the best reported bandpass $\Sigma\Delta$ modulator in Table 6-2. Also linearity performance, which is not included in the FOM, is much higher than that of the other designs. The main advantage of a bandpass modulator is the excellent image rejection performance. This does not apply to pseudo n-path designs [6.21] which have a finite image rejection ratio.

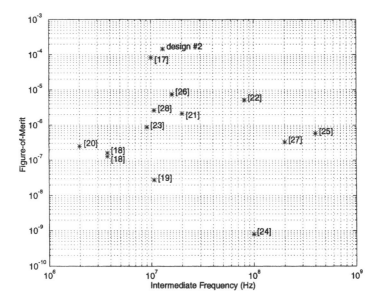

Fig. 6-2   *Figure-of-merits of IF ΣΔ modulators of Table 6-2 as a function of input IF*

6.3 Conclusions

In this chapter, the test chips that are presented in the book, have been compared with other designs from literature. The tests have been carried out with the FOMs that have been described in chapter 2. Two different benchmarks are presented. The first benchmark compares a number of lowpass ΣΔ modulators and the second one compares ΣΔ modulators for IF A/D conversion. It was shown by the first benchmark that the FOM of the continuous-time lowpass ΣΔ modulator of this book is comparable with high performance audio ΣΔ modulators from literature. Based on this design, the second test chip includes a mixer for frequency translation of an IF input to DC, or a low offset frequency. From the second benchmark it becomes clear that the combined design of a lowpass ΣΔ modulator and mixer is much more power efficient than bandpass ΣΔ modulators, performing direct IF digitalization. In the FOM of the second benchmark, linearity was not included. Table 6-2 shows that the IP3 of the second test chip is better than the other reported values.

The results of the benchmark are in agreement with the conclusions of chapter 2, where it was elaborated that it is more difficult to achieve good linearity, dynamic range and power efficiency at high frequencies than at low frequencies. With the use of dynamic element matching, high accuracy can be achieved in analog circuits as well. With the techniques that have been presented in this book, a high performance and power efficient IF A/D converter is obtained. Such an A/D converter is a key element that is required for highly integrated receivers.

References

[6.1] Marques, A.M., V. Peluso, M.S.J. Steyaert, W. Sansen, "A 15-b Resolution 2-MHz Nyquist Rate $\Delta\Sigma$ ADC in a 1-μm CMOS Technology," *IEEE J. Solid-State Circuits*, vol. 33, pp. 1065-1075, Jul. 1998.

[6.2] Thanh, C.K., S.H. Lewis, P.J. Hurst, "A Second-Order Double-Sampled Delta-Sigma Modulator Using Individual-Level Averaging," *IEEE J. Solid-State Circuits*, vol. 32, pp. 1269-1273, Aug. 1997.

[6.3] Chen, F. B.H. Leung, "A High Resolution Multibit Sigma-Delta Modulator with Individual Level Averaging," *IEEE J. Solid-State Circuits*, vol. 30, pp. 453-460, Apr. 1995.

[6.4] Burmas, T.V., K.C. Dyer, P.J. Hurst, S.H. Lewis, "A Second-Order Double-Sampled Delta-Sigma Modulator Using Additive-Error Switching," *IEEE J. Solid-State Circuits*, vol. 31, pp. 284-293, Mar. 1996.

[6.5] Zwan, E.J. van der, E.C. Dijkmans, "A 0.2-mW CMOS $\Sigma\Delta$ Modulator for Speech Coding with 80 dB Dynamic Range," *IEEE J. Solid-State Circuits*, vol. 31, pp. 1873-1880, Dec. 1996.

[6.6] Peluso, V., P. Vancorenland, A. Marques, M. Steyaert, W. Sansen, "A 900 mV 40 μW Switched Opamp $\Delta\Sigma$ Modulator with 77 dB Dynamic Range," *ISSCC Dig. Tech. Papers*, pp. 68-69, Feb. 1998.

[6.7] Coban, A.L., P.E. Allen, "A 1.5 V 1.0 mW Audio $\Delta\Sigma$ Modulator with 98 dB Dynamic Range," *ISSCC Dig. Tech. Papers*, pp. 50-51, Feb. 1999.

[6.8] Geerts, Y., M. Steyaert, W. Sansen, "A 2.5 Msample/s Multi-bit $\Delta\Sigma$ CMOS ADC with 95 dB SNR," *ISSCC Dig. Tech. Papers*, pp. 336-337, Feb. 2000.

[6.9] Fujimori, I. L. Longo, A. Hairapetian, K. Seiyama, S. Kosic, J. Cao, S. Chan, "A 90 dB SNR, 2.5 MHz Output Rate ADC using cascaded Multibit $\Delta\Sigma$ Modulation at 8x Oversampling Ratio," *ISSCC Dig. Tech. Papers*, pp. 338-339, Feb. 2000.

[6.10] Geerts, Y., A. Marques, M. Steyaert, W. Sansen, "A 3.3 V 15-bit Delta-Sigma ADC with a Signal Bandwidth of 1.1 MHz for ADSL-Applications," *Proc. of ESSCIRC*, pp. 168-171, Sep. 1998.

[6.11] Rabii, S., B.A. Wooley, "A 1.8-V Digital-Audio Sigma-Delta Modulator in 0.8-μm CMOS," *IEEE J. Solid-State Circuits*, vol. 32, pp. 783-796, Jun. 1997.

[6.12] Williams, L.A., B.A. Wooley, "A Third-Order Sigma-Delta Modulator with Extended Dynamic Range," *IEEE J. Solid-State Circuits*, vol. 29, pp. 193-202, Mar. 1994.

[6.13] Yin, G., F. Stubbe, W. Sansen, "A 16-b 320-kHz CMOS A/D Converter Using Two-Stage Third-Order ΣΔ Noise Shaping," *IEEE J. Solid-State Circuits*, vol. 28, pp. 640-647, Jun. 1993.

[6.14] Zwan, E.J. van der, "A 2.3 mW CMOS ΣΔ Modulator for Audio Applications," *ISSCC Dig. Tech. Papers*, pp. 220-221, Feb. 1997.

[6.15] Grilo, J., E. MacRobbie, R. Halim, G. Temes, "A 1.8 V 94 dB Dynamic Range ΔΣ Modulator for Voice Applications," *ISSCC Dig. Tech. Papers*, pp. 230-231, Feb. 1996.

[6.16] Rabii, S., B.A. Wooley, "A 1.8 V, 5.4 mW, Digital-Audio ΣΔ Modulator in 1.2 μm CMOS," *ISSCC Dig. Tech. Papers*, pp. 228-229, Feb. 1996.

[6.17] CHen, F., and B. Leung, "A 0.25 mW Low-Pass Passive Sigma-Delta Modulator with Built-In Mixer for a 10-MHz IF Input," *IEEE J. Solid-State Circuits*, vol. 32, pp. 774-782, Jun. 1997.

[6.18] Jantzi, S.A., K.W. Martin, and A.S. Sedra, "Quadrature Bandpass ΔΣ Modulation for Digital Radio," *IEEE J. Solid-State Circuits*, vol. 32, pp. 1935-1950, Dec. 1997.

[6.19] Singor, F.W. and W.M. Snelgrove, "Switched-Capacitor Bandpass Delta-Sigma A/D Modulation at 10.7 MHz," *IEEE J. Solid-State Circuits*, vol. 30, pp. 184-192, March 1995.

[6.20] Song, B.S., "A Fourth-Order Bandpass Delta-Sigma Modulator with Reduced Number of Op Amps," *IEEE J. Solid-State Circuits*, vol. 30, pp. 1309-1315, Dec. 1995.

[6.21] Ong, A.K. and B.A. Wooley, "A Two-Path Bandpass ΣΔ Modulator for Digital IF Extraction at 20 MHz," *IEEE J. Solid-State Circuits*, vol. 32, pp. 1920-1934, Dec. 1997.

[6.22] Hairapetian, A., "An 81 MHz IF Receiver in CMOS," *ISSCC Dig. Tech. Papers*, pp. 56-57, Feb. 1996.

[6.23] Engelen, J. van, R. van de Plassche, E. Stikvoort, A. Venes, "A 6th-Order Continuous-Time Bandpass ΣΔ Modulator for Digital Radio IF," *ISSCC Dig. Tech. Papers*, pp. 56-57, Feb. 1999.

[6.24] Tao, H., J.M. Khoury, "A 100 MHz IF, 400 MSample/S CMOS Direct-Conversion Bandpass ΣΔ Modulator," *ISSCC Dig. Tech. Papers*, pp. 60-61, Feb. 1999.

[6.25] Namdar, A., B.H. Leung, "A 400 MHz 12 b 18 mW IF Digitizer with Mixer Inside a ΣΔ Modulator Loop," *ISSCC Dig. Tech. Papers*, pp. 62-63, Feb. 1999.

[6.26] Tabatabaei, A., K. Kaviani, B. Wooley, "A Two-Path Bandpass $\Sigma\Delta$ Modulator with Extended Noise Shaping," *ISSCC Dig. Tech. Papers*, pp. 342-343, Feb. 2000.

[6.27] Maurino, R., P. Mole, "A 200 MHz IF, 11 bit, 4th order Band-Pass $\Delta\Sigma$ ADC in SiGe," *Proc. of ESSCIRC*, pp. 74-77, Sep. 1999.

[6.28] Tonietto, D., P. Cusinato, F. Stefani, A. Baschirotto, "A 3.3 V CMOS 10.7 MHz 6th-order bandpass $\Sigma\Delta$ modulator with 78 dB dynamic range," *Proc. of ESSCIRC*, pp. 78-81, Sep. 1999.

Index

A

A/D converter 97
accumulator 3
aliasing 44
AM/FM radio 22, 129
analog signals 3
anti-aliasing filter 11
autocorrelation 35

B

bandpass A/D conversion 13
base station 13
benchmark 144
Butterworth filter 56

C

ceramic filter 11
channel filter 11
characteristic equation 48, 54
charge injection 125
chip micrograph 99, 135
clock feedthrough 88
comparator 90
complex data-controlled DEM 117
complex filter 66
complex integrator 64
cross-modulation distortion 17

D

D/A converter 97, 133
decimation filter 4
degeneration 96
delay-locked loop 133
delta modulator 29

DEM circuit 133
DEM control signal 114
digital receiver 14
digital signal processor 9
distortion 80, 93, 100
dither 35
dynamic element matching 15, 111
dynamic range 16, 102

F

feedback compensation 60
feedforward compensation 54
figure-of-merit 21, 144

G

GSM band 22

H

harmonic distortion 17, 37
high frequency stability 53
higher-order filters 53

I

IF digitizing 13
IF-to-baseband $\Sigma\Delta$ modulator 73
image interference 19, 67, 113
image rejection 20, 110, 135
integrator 3, 30, 90
interference signals 9
intermediate frequency 11
intermodulation distortion 17, 105, 145
intermodulation intercept point 17
intersymbol interference 41
inverse Chebyshev 59

J

jitter 43, 105

L

large signal stability 51
latch 97
limit cycle 49
limiter 52
linear region 81
local oscillator 11, 82
low-IF 12, 129

M

matching 111
measurement setup 100
mismatch 123
mixer 11, 80, 131
mixer inside the sigma-delta loop 74
mixer outside the sigma-delta loop 77

N

noise transfer function 32
noise-shaping 3, 30
non-linear transconductance 93
non-linearity 17
non-overlapping clock 83, 98, 103
numerical model 5
Nyquist frequency 3

O

offset 33, 87
opamp 92
OTA 96
overlapping clock 83
overload 52, 102
oversampling 3, 30
oversampling ratio 31

P

parasitic delay 111
passive mixer 79, 80
peak SNR 16
peaking 63
phase noise 43
phase uncertainty 47

power consumption 22, 144
pseudorandom 115

Q

quadrature LO clock 110, 132
quadrature mixing 11
quadrature $\Sigma\Delta$ modulator 110
quadrature signals 110
quantization error 3, 31
quantization noise 31
quantizer 47

R

RC integrator 90
resolution 3, 16
resonator 59, 129
return-to-zero 42, 76
root locus 48

S

sampling rate 3
SAW filter 11
selectivity 9
self-mixing 88, 126
sensitivity 9
signal bandwidth 3
signal transfer function 32
signal-to-noise ratio 16
signal-to-noise-and-distortion ratio 17
small signal stability 46
spurious-free dynamic range 17
stability 46
superheterodyne receiver 11
switch impedance 80

T

thermal noise 95
third harmonic distortion 39, 95
tones 33, 102
transconductance-C integrator 96

W

waveform asymmetry 41
white noise 30

Z

zero-IF 12
zero-order hold 48

Lightning Source UK Ltd.
Milton Keynes UK
UKOW06n2056140415

249627UK00001B/9/P